Activated Sludge Separation Problems

Scientific and Technical Report No.16

Activated Sludge Separation Problems

Theory, Control Measures, Practical Experience

IWA SPECIALIST GROUP ON
ACTIVATED SLUDGE POPULATION DYNAMICS

Edited by Valter Tandoi, David Jenkins
and Jiri Wanner

IWA
Publishing

LONDON • SEATTLE

Published by IWA Publishing, Alliance House, 12 Caxton Street, London SW1H 0QS, UK
Telephone: +44 (0) 20 7654 5500; Fax: +44 (0) 20 7654 5555; Email: publications@iwap.co.uk
Web: www.iwapublishing.com

First published 2006

Index prepared by Indexing Specialists, Hove, UK.

British Library Cataloguing in Publication Data
A CIP catalogue record for this book is available from the British Library

Library of Congress Cataloging- in-Publication Data
A catalog record for this book is available from the Library of Congress

ISBN 1900222841
ISBN 9781900222846

Contents

Abbreviations

AS	Activated sludge
$A_2O^{®}$	Aerobic/Aerobic
Anammox®	Anaerobic ammonia oxydation
BCFS®	Biological/Chemical Phosphorus and Nitrogen removal
BD	Biodenitro
BNR	Biological Nutrients Removal
EBPR	Enhanced Biological Phosphorus Removal
FI	Filament index
GALO	*Gordonia amarae* like organisms
HRT	Hydraulic retention time
LCFAs	Long chain fatty acids
MCRT	Mean cell retentation time (equal to sludge age)
NALO	*Nocardia amarae* like organisms
N/DN	Nitrification/denitrification
OVR	Oxygen uptake rate
PE	Population equivalent
PP	Polyphosphate
PHA	Polyhydroxyalkanoate
PHB	Polyhydroxybutyrate
PHV	Polyhydroxyvalerate
16S rDNA	16S ribosomial RNA gene (present in DNA)
RAS	Return activated sludge
SCFA	Short chain fatty acids
SFN	Specific filament index
SSVI	Stirred sludge volume index
UCT	University of Cape Town
VFAs	Volatile fatty acids
WWTP	Wastewater treatment plant

List of contributors

Andreas Andreadakis	Athens NTU, Water Res. Dept, Athens, Greece
Kjær K. Andreasen	DAKA a.m.b.a, Denmark
Linda L. Blackall	Advanced Wastewater Management Centre, QLD University, Brisbane, (QLD), Australia
Glen T. Daigger	CH2M HILL, Englewood (CO), USA
Davide Dionisi	University La Sapienza, Chemistry Dep., Rome, Italy.
Dick H. Eikelboom	ASIS, Zutphen, The Netherlands
Richard Foot	Wessex Water, Dorset, UK
David Jenkins	David Jenkins, UCB, Berkeley (CA), Usa
Olga Krhutkova	Prague Institute of Chem. Technol. & Env. Engin., Praha, Czech Republic
Hilde Lemmer	Bavarian State Office for Water Management, Munich, Germany
Caterina Levantesi	IRSA-CNR, Rome, Italy
Kenneth C. Lindrea	La Trobe University, BRC, Bendigo, (Vic). Australia
Mauro Majone	University La Sapienza, Chemistry Dep., Rome, Italy
Norbert Matsché	University of Technology, Wien, Austria
Christine A. McKenzie	La Trobe University, BRC, Bendigo, (Vic). Australia
Takashi Mino	Univ. of Tokyo, Urban Engin., Dep., Tokyo, Japan
Anthony R. Pitman	Johannesburg Water, Marshalltown, South Africa
Roger Pujol	Lyonnaise des Eaux, Paris, France
Mike Robinson	Wessex Water, Dorset, UK
Ivheta Ruzickova	Prague Institute of Chem. Technol. & Env. Engin., Praha, Czech Republic
Roberto Ramadori	IRSA-CNR, Rome, Italy
Simona Rossetti	IRSA-CNR, Rome, Italy
Elisabeth Seviour	La Trobe University, BRC, Bendigo, (Vic). Australia
Robert J. Seviour	LaTrobe University-P.O. Box 199-Bendigo, Vic 3550-Australia
Valter Tandoi	IRSA-CNR, Rome, Italy
Maria C. Tomei	IRSA-CNR, Rome, Italy
Jiri Wanner	Prague Institute of Chem. Technol. & Env. Engin., Praha, Czech Republic

1

Wastewater characterization

M.C. Tomei

1.1 INTRODUCTION

In the past, wastewater characterization data were limited to BOD_5 and COD, suspended solids and ammonia concentrations. In the last few decades model development has required wastewater characterization in order to better describe complex biological processes system such as activated sludge.

The first simple models based on Monod kinetics that considered one carbonaceous substrate and one biomass as homogeneous substances were inadequate to explain the bacterial selection mechanisms that determine the equilibrium conditions in the reactors and the plant response under dynamic conditions, consequently it was necessary to introduce a more detailed substrate characterization. The first model to include the distinction between soluble and suspended carbonaceous substrate (COD) was the Clifft and Andrews model (1981); subsequent models (Dold *et al.*, 1980; Henze *et al.*, 1987) recognized the importance of evaluating the COD and nitrogen substrate fractions in the influent not only in terms of the physical state but also in terms of biodegradability. This approach, in fact, allows to better take into account the role of the colloidal fraction that includes components readily and slowly biodegradable. The COD characterization as a function of biodegradability is also the most effective in relation to the bulking phenomena, being the kinetic competition between flocforming and filamentous bacteria for readily biodegradable COD, one of the most important selective mechanisms causing the predominance of one of the two species in the sludge biocenosis (Kappeler and Gujer, 1994).

Detailed and structured knowledge of wastewater composition is also necessary for understanding the bulking and foaming problems and for finding proper control measures.

 Activated Sludge Separation Problems: Theory, control measures and practical experience edited by V. Tandoi, D. Jenkins And J. Wanner. ISBN: 1900222841. Published by IWA Publishing, London, UK.

In the following paragraphs the different approaches to wastewater characterization will be presented.

1.2 GROSS PARAMETERS

Sewage is a complex mixture of inorganic and organic substances. The chemical and physical nature of municipal wastewater can depend on variable industrial contributions.

Average waste water characteristics are difficult to determine because of its temporal variability.

The simplest method of characterizing municipal wastewater is determination of “gross parameters” such as BOD, COD, TSS, TKN and TP concentrations. Table 1.1 shows annual pollution loads for different countries. Tables 1.2 and 1.3 show pollutant concentration data for settled and raw municipal wastewater.

Table 1.1. Annual pollution load in different countries (from Henze *et al.*, 1995b)

COUNTRY	PARAMETER (kg $pers^{-1} \cdot yr^{-1}$)				
	BOD	TSS	N_T	P_T	Detergents
Brazil	20-25	20-25	3-5	0.6-1	0.5-1
Denmark	20-25	30-35	5-7	1.5-2	0.8-1.2
Egypt	10-13	15-25	3-5	0.4-0.6	0.3-0.5
Germany	20-25	30-35	4-6	1.2-1.6	0.7-1
India	10-15	-	-	-	-
Italy	18-22	20-30	3-5	0.6-1	0.5-1
Sweden	25-30	30-35	4-6	0.8-1.2	0.7-1
Turkey	10-15	15-25	3-5	0.4-0.6	0.3-0.5
Uganda	20-25	15-20	3-5	0.4-06	-
USA	30-35	30-35	5-7	1.5-2	0.8-1.2

Table 1.2. Typical pollutant concentration of settled municipal wastewater in different countries (Henze *et al.*, 1987; Ekama and Marais, 1984). Data for Italy are referring to Regione Emilia Romagna (Regione Emilia Romagna ARPA E.R., 2002)

PARAMETER (mg l^{-1})	COUNTRY				
	Denmark	Italy	Hungary	South Africa	Switzerland
BOD_5	100-350	220	-	150-300	-
COD	210-740	620	350	300-600	220
SS	120-450	300	-	150-300	-
TKN	20-80	-	58	30-70	27
Ammonia N	12-50	42	30	24-56	10
Total P	4-23	7	-	6-15	-

Table 1.3. Typical pollutant concentration of raw municipal wastewater in USA and UK (data from Gray,1992)

PARAMETER (mg l^{-1})	COUNTRY	
	USA	U.K
BOD	250	326
COD	500	650
TOC	250	173
SS	220	127
Total N	40	66
Ammonia N	25	47
Organic N	25	19
Total P	12	-

1.3 PHYSICAL PROPERTIES

The physical properties of sewage are odour, temperature, pH and solids distribution.

The typical musty smell of sewage is produced by the two main groups of nitrogen and sulphur compounds and to a minor extent chlorine and phenol compounds. Amines, ammonia, diamines and skatole are included in the first group. The second is comprised of hydrogen sulphide, mercaptans, organic sulphide and sulphur dioxide while the third one of chlorine and chlorophenol.

Wastewater temperature depends on latitude, with values ranging from 8° C during winter in Northern Europe to 28-30 °C in the summer in Asiatic countries (Gray, 1992). The pH value is generally in the range of 6.7-8.2. The suspended particle size varies from 1 μm to distinguishable matter: some suspended solids are settleable, but the majority are colloidal particles that are not removed by settling. A simple classification based on size (Painter, 1971) distinguishes solid particles as settleable (>100 μm, about 50%), supracolloidal (1-100 μm, 30-70 %) and colloidal (1nm-1 μm) the remaining 17-20 %.

A more detailed classification of the solids fractions is presented in Table 1.4.

Table 1.4. Typical solid fraction and related COD in sewage (modified from Gray, 1992)

Solid fraction	Total solids		COD	
	mg l^{-1}	%	mg l^{-1}	%
Settleable	74	15	120	29
Supracolloidal	57	11	87	21
Colloidal	31	6	43	10
Soluble	351	68	168	40
Total	513	-	418	-

1.4 ORGANIC MATTER

Approximately 75% of the total organic carbon is made up of: carbohydrates, fats, proteins, amino-acids and volatile fatty acids (Gray 1992). The remainder are a myriad of material including surfactants, vitamins, antibiotics, purines, hormonal compounds, pesticides, hydrocarbons, pigments and other synthetic compounds. Some organic compounds are difficult to biodegrade. Sewage composition depends on sewer characteristics. Long anaerobic retention times favour conversion of carbohydrates to volatile fatty acids and the rapid transformation of urea in ammonia as well as solubilization of particulates.

Table 1.5 shows the distribution of organic carbon compound groups in solution and in suspension.

The carbohydrates are mainly constituted by glucose, sucrose and lactose with smaller proportions of galactose, fructose, xylose and arabinose and have a typical concentration interval of 50 -120 mg l^{-1}.

Table 1.5. Typical distribution of organic constituents in municipal wastewater (modified from Painter, 1983)

Constituent	In solution		In suspension	
	Concentration (mgl^{-1})	C/C Total (%)	Concentration (mgl^{-1})	C/C Total (%)
Fats	-	-	140	50
Carbohydrates	70	31.3	34	6.4
Free and bound amino acids	18	10.7	42	10
Volatile acids	25	11.3	12.5^1	2.3^1
Non-volatile acids	34	15.2		
Detergents (ABS)	17	11.2	5.9	1.8
Uric acid	1	0.5	-	-
Creatine	6	3.9	-	-
Amino sugars	-	-	1.7	0.3
Amides	-	-	2.7	0.6

1) Sum of volatile and non-volatile acids

In wastewater characterization the word fats is a general term used to indicate not only fats but also oils and waxes present in wastewater. They mainly originate from food preparation and to a minor extent from excreta, are weakly soluble, and are found essentially in the suspended fraction of wastewater at a concentration ranging from 40-150 mg l^{-1}. Fats are characterized by low biodegradation kinetics due to the fact that they need to be hydrolyzed prior to being metabolized by microorganisms. Hydrolysis products are fatty acids that increase the content of free fatty acids present in the sewage. A fraction of 70-90 % of total fatty acids is constituted by palmitic, stearic and oleic acids.

Among volatile acids detected in sewage the most important is acetic acid, whose concentration is in the range of 6-37 mg l^{-1}. It plays an important role in the storage phenomena and consequently in microbial selection. Other volatile acids present in significant quantities are propionic, butyric and valeric that, together with acetic constitute 75-90% of the total acidity present in municipal wastewater. Non volatile acids such as glutaric, glycolic, lactic, citric and benzoic acid are present to a minor extent with concentrations ranging from 0.1-1 mg l^{-1}.

Proteins are also a source of carbon but their contribution is most significant for organic nitrogen that constitutes about 16% of their structure. The breakdown of proteins by bacterial action gives rise to the formation of free amino acids, fatty acids, nitrogenous compounds, phosphates and sulphides. Concentrations for free and bonded amino acids (peptides and proteins) are <5 mgN l^{-1} and in the range of 4-15 mgN l^{-1}, respectively.

Urea is the major source of nitrogen (2-16 mgN l^{-1}); depending on the retention time in the sewer system, a significant part of urea is converted to ammonia both in anaerobic and aerobic conditions.

The remaining fraction of biodegradable matter not included in the above mentioned categories are compounds resulting from the combination of carbohydrates, proteins and fats such as lipoproteins and nucleoproteins. Lignins and cellulose constitute a slightly degradable fraction, while the organic unbiodegradable fraction is 20 - 40 % of the total organic matter.

Organisms deriving from human excreta, soil and water including viruses, bacteria, fungi and protozoans are also present in sewage. A considerable fraction of them is pathogenic and has to be removed in wastewater treatment plants.

1.5 INORGANIC MATTER

The content of inorganic matter in sewage is determined by two main factors: the characteristics in terms of the hardness of the water supplied and the nature of the pollutants. In hard wastewater the predominant ions are calcium, sodium and chloride. Other components detected in significant amounts are potassium, magnesium, sulphur compounds (mainly sulphates), phosphates, bicarbonates and ammonia.

Nitrogen and phosphorus are essentially the most important nutrients for micro-organism growth. Inorganic nitrogen that gives the majority of the total nitrogen is constituted by ammonia which is found in solution as ammonium (NH_4^+) or ammonia (NH_3) depending on the pH. The oxidized nitrogen forms are present in sewage only in very small amounts.

Phosphorus is found in three different forms, ortophosphate, polyphosphate and organic phosphate. Ortophosphate is the predominant form and is directly assimilable by organisms. Both organic phosphate and polyphosphates must be transformed to ortophosphate prior to being utilized both in the plant and in the receiving waters, as a result the orthophosphate concentration in the plant effluent is an important parameter in evaluating the expected impact on the receiving water.

The average data for nitrogen and phosphorous in municipal wastewater are shown in Tables 1.2 and 1.3.

Other elements necessary for biological growth (but only in traces) are sulphur and metals that are usually present in sufficient (or excessive) amounts in sewage. Excessive amounts of metals can exert a toxic or inhibitory effect on the biological processes and are a problem in sludge treatment and disposal.

The range of concentration of metals in sewage is reported in Table 1.6.

Table 1.6. Typical concentration of heavy metals in municipal wastewater

Heavy metal ($mg\ m^{-3}$)	Gould (1976)	Henze *et al.* (1995b)
Al	-	250-1000
As	-	1-5
Cd	20	1-4
Cr	400	10-40
Co	<20	0.5-2
Cu	880	30-100
Fe	800	400-1500
Pb	250	25-80
Mn	200	40-150
Hg	-	1-3
Ni	<20	10-40
Ag	-	3-10
Zn	500	80-300

1.6 BIODEGRADABILITY

1.6.1 COD, Nitrogen fractionation

In relation to bulking and foaming understanding and control, the detailed wastewater characterization in terms of compound groups can be simplified by distinguishing carbon and nitrogen fractions as the function of biodegradability. This approach has been proposed by general activated sludge models (Dold *et al.*, 1980; Henze *et al.*, 1987) and is reported in detail in the IAWQ Scientific and Technical Reports No.1 (Henze *et al.*, 1987) and No. 2 (Henze *et al.*, 1995a).

According to the nomenclature proposed in the Activated sludge model No.1, the carbonaceous substrate, expressed as COD, is constituted by the following fractions:

- readily biodegradable substrate (S_S)
- slowly biodegradable substrate (X_S)
- inert suspended organic matter (X_I)
- inert soluble organic matter (S_I)

The readily biodegradable substrate (that can be considered soluble in first approximation) consists of simple molecules directly metabolizable by heterotrophic bacteria for their maintenance and growth. VFA (Volatile Fatty Acids), lower alcohols, lower amino acids and simple carbohydrates (Henze *et al.*, 1994) are the main groups included in this fraction.

Slowly biodegradable substrate (that in first approximation can be considered particulate) consists of more complex compounds that need to be hydrolyzed prior to be utilized by the micro-organisms. The slow biodegradability is a consequence of the lower kinetics of the hydrolysis process compared to the utilization kinetics, therefore the hydrolysis is the rate limiting step of the whole biodegradation process. Both readily biodegradable substrates present in wastewater and the hydrolysis products support the growth of certain type of filamentous bacteria.

Inert suspended organic matter is constituted by the inorganic fraction of the suspended solids and part of the suspended organic matter that is not biodegradable. In this fraction compounds very slowly biodegradable (so practically unbiodegradable under the usual operating conditions) such as lignin and cellulose are also included.

Inert soluble organic compounds are presumably constituted by xenobiotic compounds that are unbiodegradable and are found integrally in the effluent.

In the Activated Sludge Model No. 2 (Henze *et al.*, 1995a), that also includes the biological phosphorus removal, a further subdivision has been considered: the S_S is constituted by the sum of two components: fermentable readily biodegradable organic substances S_F and S_A, fermentation products assimilable to acetate.

The main distinction as regards to the nitrogenous substrate is between ammonia and organic components. For the latter the same approach as for COD is considered. Nitrite and nitrate are present to a minor extent. Consequently, nitrogen in municipal wastewater can be divided into the following fractions:

S_{NH4} = ammonium nitrogen
S_{NO3} = nitrite plus nitrate nitrogen
S_{ND} = soluble biodegradable organic nitrogen (assimilable to readily biodegradable N)
X_{NS} = particulate biodegradable organic nitrogen (assimilable to slowly biodegradable N)
S_{NI} = soluble inert organic nitrogen
X_{NI} = particulate inert organic nitrogen

Table 1.7. COD and Nitrogen fractions in settled municipal wastewater in different countries (data from Henze *et al.*, 1987 and Ekama & Marais, 1984)

PARAMETER	UNITS	COUNTRY			
		Denmark	Switzerland	Hungary	South Africa
S_s	mg COD l^{-1}	125	70	100	68-135
S_I	mg COD l^{-1}	40	25	30	31-62
X_s	mg COD l^{-1}	250	100	150	210-420
X_I	mg COD l^{-1}	100	25	70	15-30
S_{ND}	mg N l^{-1}	8	5	10	3.6-9
X_{ND}	mg N l^{-1}	10	10	15	2.4-5
S_{NH}	mg NH_3^-·N l^{-1}	30	10	30	24-56
S_{NI}	mg N l^{-1}	2	2	3	1.5-3.5
S_{NO}	mg NO_3·N l^{-1}	0.5	1	1	-

The soluble inert compounds are usually negligible while the particulate substances are associated with the suspended carbonaceous substrate. The soluble oxidizable fraction is constituted essentially by ammonia and its salts and, to a minor extent, by soluble organic nitrogen, while the biodegradable particulate, likewise the particulate biodegradable COD, is hydrolyzed to soluble organic nitrogen which in turn is converted to ammonia.

Phosphorus is also found in dissolved and suspended phases: the dissolved fraction includes inorganic orthophosphate (being the prevalent fraction), inorganic polyphosphate and organic phosphorus, while the suspended fraction is constituted by organic phosphorus associated with suspended COD.

Figure 1.1 shows the carbonaceous substrate fractionation in terms of groups of compounds and related biodegradability in municipal wastewater while data available for different countries are reported in Table 1.7. The inappropriate ratio between COD (S_S-X_S) and N and P concentration result in bulking by certain types of filaments (see Section 5.2.2).

1.6.2 Evaluation methods

1.6.2.1 Carbonaceous substrate (COD)

S_I The dissolved inert fraction can be evaluated by utilizing a simple bench scale apparatus; a small reactor where the wastewater is mixed with activated sludge at the same F/M (food/microorganisms) ratio as in the full-scale plant. The residual COD in the effluent, measured after a contact time of 20-30 days, gives the S_I value. Control parallel tests without substrate are advisable (Henze *et al.*, 1995b).

S_S Two methods are actually available for readily biodegradable COD evaluation: the respiration (or biological) method (Ekama *et al.*, 1986) and the flocculation method (Mamais *et al.*, 1993). The first one is accurate and reliable but time consuming and difficult to utilize routinely. A bench scale plant consisting of an aeration basin and a secondary settler is required for the measurement. This biological plant is operated at 2.5 days MCRT in a semicontinuous mode with intermittent feed (12 hours feed on, 12 hours feed off). The S_S value is determined on the basis of OUR (Oxygen Uptake Rate) step variation (a sudden decrease) detected in the reactor after the feed period.

The S_S evaluation requires the knowledge of the growth yield coefficient for the heterotrophic biomass that the Authors assume equal to 0.67 mg biomass(COD)/mg substrate(COD). All the details for the experimental procedure are reported in (Ekama *et al.*, 1986).

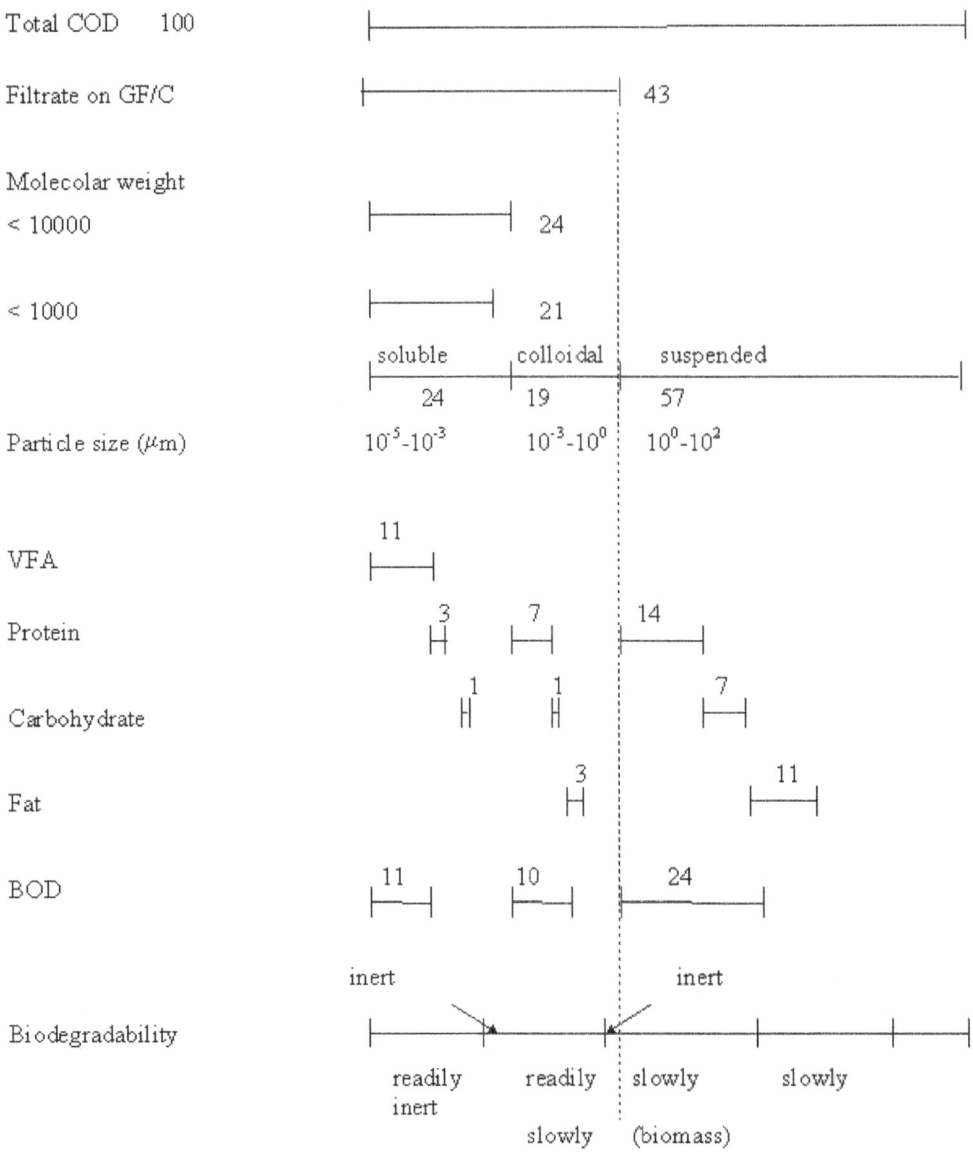

Figure 1.1. Carbonaceous substrate fractionation and biodegradability in municipal wastewater (modified from Henze *et al.*, 1994)

The flocculation method, compared to the biological, is more rapid and easier to use in the day-to-day management control and operation of a treatment plant. The principle of this method is based on the experimental evidence that membrane filtration of a sample that has been previously flocculated (in this case by precipitating $Zn(OH)_2$ at pH 10.5) will produce a filtrate containing only truly soluble organic matter due to the fact that the colloidal particles, which usually pass through the filter, are removed during the flocculation step. The measurement is carried out by the following sequence: the sample is flocculated by adding zinc sulphate, then the pH is adjusted to 10.5 by dosing sodium hydroxide and, finally, the clear supernatant (obtained after a quiescence time of a few minutes) is filtered using a 0.45 μm membrane filter. The COD of the filtrate (COD_{sol}) is related to the S_S by the following:

$$S_S = COD_{sol} - S_I$$

Consequently, an estimation of S_I is required to evaluate S_S.

All the details of the experimental procedure are reported in Mamais *et al*, 1993.

At present the biological determination is believed to be the safest (Henze *et al.*, 1995b) but due to the complexity of the measurement a good compromise could be to employ the physical-chemical determination on the day to day practice and periodically verify the reliability of the obtained values by utilizing the respirometric test as a control.

S_A This fraction is mainly constituted by acetic acid, so a reasonable estimation can be obtained by determining the acetic acid concentration utilizing the usual chromatographic methods (gas chromatography and ion chromatography).

X_I The procedure for estimating the inert suspended organic fraction is quite complex. Experimental data obtained from continuous lab or pilot scale plants at different sludge age are required. The sludge production data are analyzed with a mathematical model of the activated sludge process and the X_I concentration is estimated by calibrating the model in order to obtain the best fitting of the sludge production data. Details on this procedure are reported in Gujer and Henze (1991).

X_S Finally, the slowly biodegradable organic fraction is determined as the difference between the total COD (COD_T) and the other fractions:

$$X_S = COD_T - S_I - S_S - S_A - X_I$$

1.6.2.2 Nitrogen and phosphorus

Nitrogen fractions can be evaluated by conventional analysis methods for Kjeldahl nitrogen (TKN including ammonia and organic nitrogen), nitrite, nitrate and ammonium (APHA, 1992).

An approximate estimation of the organic nitrogen fractions can also be obtained from the COD values considering that the nitrogen/COD ratio in the various organic fractions for different domestic wastewaters varies in a quite narrow range 0.04 - 0.08 mgN/mgCOD (Henze *et al.*, 1995b).

Phosphorus fractions determination can be easily performed with the usual well-tested methods (APHA, 1998).

1.7 REFERENCES

APHA (1998). Standard Methods for the Examination of Water and Wastewater, 20th edition.

Clifft R. C., Andrews J.F. (1981). Predicting the dynamics of oxygen utilization in the activated sludge process. *J Wat Pollut Control Fed,* 53, 1219- 1232.

Dold P.L., Ekama G.A., Marais G.v.R. (1980). A general model for the activated sludge process. *Prog Wat Tech,* **12,** 47-77.

Ekama G.A., Marais G.v.R. (1984). Nature of municipal wastewaters. In *Theory design and operation of nutrient removal activated sludge processes.Chap. 2,* Water Research Commission, Pretoria South Africa.

Ekama G.A. Dold P.L. Marais G.v.R. (1986). Procedures for determining influent COD fractions and the maximum specific growth rate of heterotrophs in activated sludge systems. *Wat Sci Technol,* **18**(6), 91-114.

Gould D.J. (1976). Ecological effects of sewage discharge to the sea. *Technical Report TR26*, Water Research Centre, Stevenage, UK.

Gray N. F. (1992). How nature deals with waste. In *Biology of wastewater treatment. Chap. 1,* Oxford University Press, UK.

Gujer W. Henze M. (1991). Activated sludge modelling and simulation. *Water Sci & Technol*, **23** (4-6), 1011-1023.

Henze M. Grady Jr. C.P.L. Gujer W. Marais G.v.R. Matsuo T. (1987). Activated Sludge Model No 1, *IAWPRC Scientific and Technical Reports No 1*, IAWPRC London, UK, 1-33.

Henze M.; Kristensen G.H., Strube R. (1994). Rate-capacity characterization of wastewater for nutrient removal processes, *Water Sci & Technol,* **29** (7), 101-107.

Henze M., Gujer W., Mino T., Matsuo T., Wentzel M.C., Marais G.v.R. (1995a). Activated Sludge Model No 2. IAWQ *Scientific and Technical Reports No 2*, IAWQ, London, UK., 1-32.

Henze M., Harremoes P., la Cour Jansen J., Arvin E. (1995b). Wastewater treatment Biological and chemical processes. Springer Verlag Berlin, Heidelberg, 1-383.

Kappeler J., Gujer W. (1994). Development of a mathematical model for “aerobic bulking”. *Water Res,* **28**, 303-310.

Mamais D., Jenkins D., Pitt P. (1993). A rapid physical chemical method for the determination of readily biodegradable soluble COD in municipal wastewaters. *Water Res*, **27**, 195- 197.

Painter H.A. (1971). Chemical physical and biological characteristics of wastes and waste effluents. In *Water and water pollution handbook 1,* Marcel Dekker, New York, 329-363.

Regione Emilia Romagna ARPA E.R. (2002). Report on Wastewater Treatment Plants.

2

The Microbiology of Activated Sludge Process

S. Rossetti, L.L. Blackall and V. Tandoi

2.1 INTRODUCTION

The Activated Sludge (AS) process is an old technology, but still the most commonly used for the treatment of domestic and industrial wastewaters. New configurations have been adopted (multistage plants for nutrient removal) and new technologies are being developed (e.g. membrane bioreactors, biofiltration, sequencing batch biofilm reactors) but all derive from the traditional AS process. Knowledge of the populations operating in these systems originated from studies of AS and have helped to operate better the process, giving scientific bases to an originally empirical approach.

Despite the wide spread use of activated sludge systems, this technology can suffer from operational dysfunction and instability in pollutant removal. This chapter contains updated information about the main bacterial populations operating in AS systems: heterotrophs, nitrifiers, denitrifiers, polyphosphate and glycogen accumulating organisms (PAO and GAO respectively). These microorganisms are important in terms of both their function and competition with filamentous bacteria, which often cause serious problems in the AS process. Knowledge of the identity and properties of these populations is crucial to properly address any strategy directed at modifying the composition of the biomass (i.e. promoting the growth of bacteria that play a positive role (such as nitrifiers and denitrifiers), or depressing the growth of problematic organisms, such as filaments. The proper identification of a bacterium is a complex procedure, and for many years the identification of many microbial population present in AS was performed using morphological criteria (i.e. cell shape,

 Activated Sludge Separation Problems: Theory, control measures and practical experience edited by V. Tandoi, D. Jenkins And J. Wanner. ISBN: 1900222841. Published by IWA Publishing, London, UK.

dimensions, Gram staining, etc.). Today, precise and rapid methods, such as Fluorescent In Situ Hybridization (FISH), are available for some bacteria, but morphological identification is still necessary and useful because information about all of the populations in AS is still incomplete.

Focus point: the identification of the microbial components of activated sludge is crucial for proper plant operation. Morphological microscopic identification represents an approximate, but still valid, evaluation tool (Eikelboom and van Buijsen, 1981 and updates; Jenkins et al., 1993 and updates, Eikelboom, 2000), but modern and precise methods of identification are quickly developing and are already available for some important bacterial components of activated sludge.

2.2 MICROORGANISMS IN ACTIVATED SLUDGE

AS is a complex microbial community composed mainly of bacteria (the real actors of the depuration process), also sometimes containing protozoa and metazoa (i.e. Rotifers, Nematodes etc.) (Table 2.1).

Bacteria: cell structure

Bacteria, the main AS components, are microorganisms, between 0.2 and 3 μm in size (but in AS are often 0.5-2 μm small), with variable cell morphology. Several distinct shapes of bacteria can be recognised, including cells with a cylindrical shape (rods), with spherical or ovoid morphology (cocci), curved rods with spiral-shaped patterns (spirillum), and tetrads (four square cells linked together).

The structure of protozoa and metazoan cells, is more complex, with a defined nucleus surrounded by a membrane and several other intracellular organelles. These organisms are called Eukaryotes, while Bacteria and Archaea are called Prokaryotes.

Figure 2.1 presents a schematic of a Gram negative and Gram positive bacterial cell. It contains the following structures:

- Cell wall, a rigid structure outside the cytoplasmic membrane, provides the cell with its structural rigidity and shape and protection against osmotic lysis.
- Cytoplasmic cell membrane, the real barrier separating inside from outside of the cell.
- Cytoplasm which contains all the complex biochemical machinery for the reproduction and the survival of the bacterial cell. The major components of the cytoplasm, other than water, are macromolecules, ribosomes (containing the ribosomal RNAs), small organic molecules serving mainly as precursors of macromolecules and various inorganic ions.
- Nucleoid, bacterial DNA in an aggregated form, not surrounded by a membrane such as in Eukaryotes.

One commonly used way to characterize bacteria is to divide them into two major groups, *Gram-positive* and *Gram-negative*. The distinction between these groups is based on differences in reactions to a particular staining procedure, known as Gram staining. These differences are due to differences in cell wall structure. Gram positive bacteria (Fig. 2.1 a) are characterised by a thicker cell wall, with about 90% of the cell wall made up of a single type of molecule, peptidoglycan. The Gram-negative cell wall has a multilayered, complex structure. It contains only about 10% peptidoglycan and also contails an outer membrane made of lipopolysaccharide.

Table 2.1 Main bacterial components present in activated sludge

Microorganism		Function	Reference
Floc-forming bacteria			
Eterotrophs	Traditional aerobic bacteria *Zoogloea, Flavobacterium, Alcaligenes, Aeromonas, Pseudomonas, Bacillus*	Organic carbon oxidation	Pike, 1972
	Denitrifiers *Pseudomonas, Bacillus, Xantomonas, Alcaligenes, Thiobacillus*	Nitrate reduction to nitrogen gas	Pike, 1972
	Polyphosphate accumulating bacteria (PAO) "Candidatus Accumulibacter phosphatis" (in *Rhodocyclus* Genus)	Biological Phosphorus removal	Hesselmann *et al.*, 1999, Crocetti *et al.*, 2000)
	G-bacteria (a morphotype tetrad forming) *Amaricoccus kaplicensis*	Carbon removal	Seviour *et al.*, 2003
	Glycogen accumulating bacteria (GAO) Candidatus "Competibacter phosphatis"	PAO's competitors	Crocetti *et al.*, 2002
Autotrophs	Ammonia oxidisers *Nitrosomonas, Nitrosospira, Nitrosococcus, Nitrosocystis*	Ammonia oxidation to nitrite	Mobarry *et al.*, 1996
	Nitrite oxidisers Nitrobacter, Nitrospina, Nitrococcus	Nitrite oxidation to nitrate	Mobarry *et al.*, 1996
	Ammonia/nitrite anaerobic utilizators (Anammox) Brocadia, Kuenenia, Scalindua	Ammonia oxidation to nitrogen by respiring nitrite	Jetten *et al.*, 2003
Filamentous bacteria	*M. parvicella, Thiothrix* sp., *Nostocoida limicola*, Type 021N, *Sphaerotilus natans*, ecc.	Organic carbon oxidation HS^- oxidation (few) P removal (not confirmed hypothesis)	Seviour *et al.*, 1999
Microfauna (Protozoa and Metazoa)	Flagellates and Ciliates Protozoa, *Rotifera, Nematoda*, etc.	Activated Sludge Indicators (Sludge Biotix Index)	Madoni (1994)

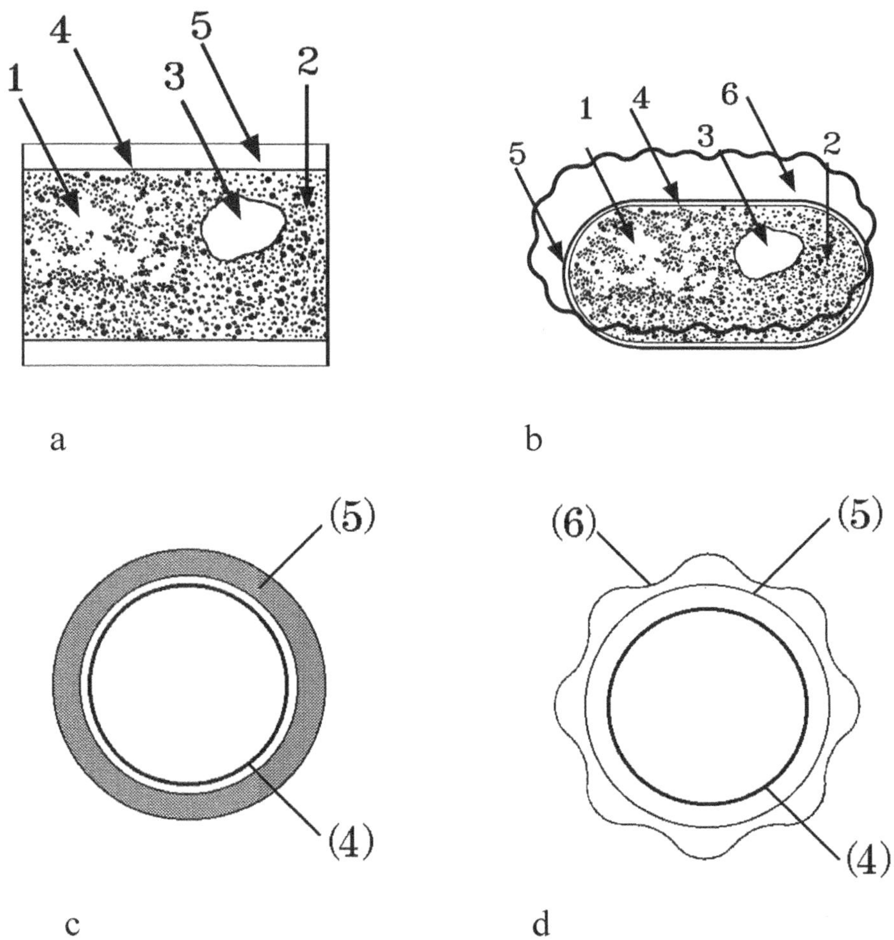

Fig. 2.1 General scheme of a Gram positive (a and c) and Gram negative (b and d) bacterial cell. 1: nucleoid; 2: cytoplasm; 3: cytoplasmic inclusion; 4: cytoplasmic membrane; 5: peptidoglycan layer; 6: outer membrane. The Gram positive cell has a ticker peptidoglycan layer and does not have the outer membrane.

Inclusions

Bacteria can contain different kinds of inclusions (such as polyhydroxyalkanoates, carbohydrates, polyphosphates, or sulphur granules) and under particular growth conditions can produce biopolymeric compounds that are excreted out of the cell (Esocellular Polymers, ECP, known generically as *capsule* material). These inclusions play an important role in AS ecology and are involved in important phenomena, such as Biological Phosphorus Removal or storage. Inclusions are a crucial part of selector operation as well.

2.3 THE IDENTIFICATION OF BACTERIA

Correct bacterial identification and enumeration is important for appropriate control of the AS process. The identification of bacteria requires the existence of a classification system (taxonomy) that depends on the determination of certain properties to verify the similarity or difference of a microorganism to other microorganisms that are already known and described. More than 18,000 bacterial species have been identified so far (but the number increases day by day), constituting probably around 1-10% of those in nature. An isolate is assigned a formal taxonomic position as a new genus or species when an accurate description of the isolate with the proposed name is published in specialized journals (such as *International Journal of Systematic and Evolutionary Bacteriology*, IJSEM). A pure culture of the organism should also be deposited in an approved culture collection, such as the American Type Culture Collection (*ATCC)* or the Deutsche Sammlung von Mikroorganismen und Zellkulturen (*DSMZ* German Collection for Microorganisms). Today, because we now know that many microorganisms are unculturable or difficult to isolate, it is common to also define the identity of an organism by molecular analysis, usually performed on highly-specialised enrichments.

Description, of new identified bacteria consist of information regarding morphology, physiology, habitat, composition of some cellular components, phenotypic and genotypic properties, and the sequence of 16S rDNA. The whole of these characteristics permits the naming of the organism, and consequently, its acceptance as a new microbiological taxon. Periodically the IJSEM publishes an approved list of bacterial names, which formalises newly proposed names and allows their inclusion in *Bergey's Manual of Systematic Bacteriology*.

In addition to this traditional and rigorous classification system, a different and more practical approach has been used to classify some relevant bacteria in AS. This second approach is based mainly on simple and easy to detect microscopic properties and, conceived by Dick Eikelboom in 1975, was essential for many years (when the quick and precise molecular methods were not available) for the elucidation of the diversity of filamentous bacteria in AS systems (Eikelboom, 1975).

Conventional taxonomy

The identification of bacteria by conventional taxonomy is based on several characteristics, such as morphology, Gram reaction, metabolic peculiarities, cell wall and lipid chemistry, DNA guanine cytosine (G+C) total bases ratio (G+C+A+T), presence of cell inclusions and storage products, temperature and pH requirements, antibiotic sensitivity, pathogenicity, and habitat, etc.. Though useful, this information does not clarify the evolutionary or phylogenetic relationships between organisms (how they developed from the previous ones during the evolution), which is a priority of new classification methods. For this reason, today a new systematic method is being used that facilitates the understanding of evolutionary relationships between all living organisms.

Molecular taxonomy

With the development of molecular techniques and their wide application to different research areas, a reorganisation of the living world is occurring. Approximately twenty years ago, living organisms were classified in five kingdoms (Animals, Plants, Fungi, Protists, and Bacteria), which were also divided into Eukaryotes (Animals, Plants, Fungi, Protists) and Prokaryotes (Bacteria) on the basis of the presence or absence of a nuclear membrane and certain structural characheristics. These two groups were thought to be independent and, because of the complexity of multicellular organisms, more biodiversity was attributed to the Eukaryotes. Recent phylogenetic analysis based on the molecular sequencing of particular molecules, has revealed a new and unsuspected classification scheme for living organisms, quite different from the previous kingdom-based classification. In particular, it was determined that the sequence of the ribonucleic acid (RNA) present in ribosomes (ribosomal RNA or rRNA) was able to serve as an evaluationary chronometer. Ribosomes are the site of protein synthesis, which is a fundamental function for the survival and the function of all cells, and are made us of a mixture of different sized RNA and proteins. The large molecules of prokaryotic ribosomal RNA, the 16S and 23S units (containing about 1500 and 2900 nucleotides, respectively) contain several regions of highly conserved sequences useful for obtaining proper sequence alignments, and also contain sufficient sequence variability in other regions to allow their utilisation as excellent evolutionary chronometers. The 5S rRNA subunit has also been utilised in this way but its small size (about 120 nucleotides) limits the information obtainable from this molecule.The rRNA sequences have been obtained from many organisms and can be organized into a phylogenetic tree, which allows the determination of the evolutiony distance between organisms (Woese, 1987).

Molecular phylogeny revealed that the five kingdoms do not represent five major evolutionary lineages, but that life on Earth has evolved along three major lineages, two of which are Prokaryotic (Bacteria and Archaea) and one is Eukaryotic (Eukarya). They represent the three domains of life, derived from the Universal Ancestor of all the living organisms. Even though Bacteria and Archaea are both prokaryotes (Figure 2.2), the root of the universal tree separates them, and surprisingly, the Archaea are closer to Eukarya than Bacteria. The Archaea branch off at a point closest the root of the Universal tree, meaning that this domain represents the least evolved whereas the Eukarya, being the farthest away from the Universal Ancestor, are the most evolved.

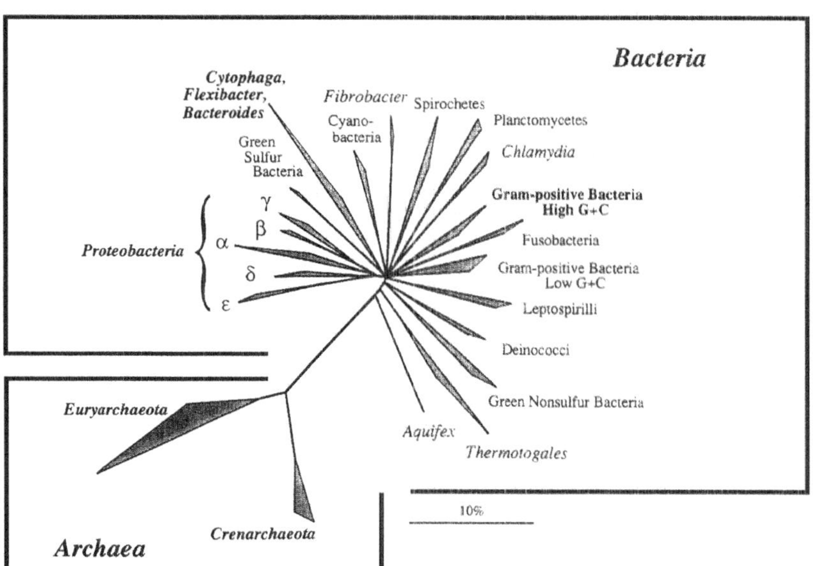

Fig. 2.2 Phylogenetic tree made by comparative analysis of 1800 ribosomial 16S rRNA sequences within "Bacteria" ed "Archaea domains (modified from Amann *et al.*, 1995)

The location of Archaea nearest the ancestor is supported by their capability to inhabit extreme environments such as those characterised by high temperatures, low pH values, and high salinity. These conditions are thought to reflect those under which life on Earth originated.

Molecular characterization of mixed biomass by molecular probes

Fluorescent In Situ Hybridization (FISH) is a widely applied technique that allows visualization of organisms in a sample, based on their phylogeny. FISH has been successfully utilized for the *in situ* monitoring of microbial population dynamics in environmental samples. FISH works through the use of molecular probes. These probes are synthesized in the laboratory and consist of short strands of nucleic acids (usually 15-30 nucleotides in length), complementary to 16S rRNA or 23S rRNA sequence regions. The oligonucleotides are usually tagged on one end with a so-called "reporter molecule", usually a fluorochrome, which allows visualization of the probe when the specimen is observed by epifluorescence microscopy. The FISH technique is described in detail in paragraph 2.13.

2.4 FILAMENTOUS BACTERIA

Filamentous organisms are characterized by their long, thread-like shape. Bacteria are the main component of filamentous organisms present in AS plants; only seldom are filamentous Eukaryotes visualised (e.g. fungi). By the systematic observation of a large number of samples from different AS treatment plants, Eikelboom (1975) described (only on morphological basis) 29 different filamentous bacteria, grouped in three clusters:

A) Bacteria ascribable to known genus and species already reported in *Bergey's Manual of Systematic Bacteriology*: i.e. *Sphaerotilus natans*, *Haliscomenobacter hydrossis*, *Beggiatoa*, *Thiothrix*, *Flexibacter*.
B) Microorganisms that, even though they have a name (i.e. *M. parvicella*, *Nostocoida limicola*) were still not present in *Bergey's Manual of Systematic Bacteriology* because of a paucity of the phenotypic information
C) Microorganisms to which it was impossible to attribute an identity and were consequently identified as a "Type" followed by a number, such as Type 021N, Type 0041, Type 0092, etc..

The filamentous species *Sphaerotilus natans* was originally thought to be the only filamentous microorganism responsible for dysfunctions in AS systems. Because of the microscopic work that has been performed since the beginning of the 1980s, it has been possible to ascertain that filamentous bulking can be caused by several different filamentous bacteria. In particular, Eikelboom's Manual (1981), a systematic guide that enables the morphological distinction of the main filamentous bacteria present in AS, has helped with such determinations.

After more than twenty four years from the publication of the first Eikelboom's Manual (1981) and twenty years from an analogous manual by Jenkins and coworkers (Jenkins *et al.*, 1984), it is necessary to underline the important contributions of these manuals that have led to the improvement of knowledge in this field.

These manuals have permitted morphological identification, monitoring, and quantification of filamentous bacteria, allowing comparisons of various corrective operations to control the overgrowth of filamentous bacteria. During recent years, pure culture studies have defined the taxonomy of some of the 29 filamentous bacteria described by Eikelboom. Several organisms that were originally thought to be a sigle species have been identified as unique species, and is some cases, adequate information has been generated to properly name the organisms. This is the case for *M. parvicella, and Nostocoida limicola* (Blackall *et al.*, 1996; Blackall *et al.*, 2000 respectively). In addition the important filament Type 021N, has been properly described and given a valid name: *Thiothrix eikelboomii* (Howarth et al., 1999).

Some of the filamentous bacteria that have not been studied are of minor importance with respect to bulking (i.e. Type 0411, Type 1702, Type 1852, Type 0211, etc.) and have not been isolated as a result of difficulties connected in picking organisms present in very low numbers. Recently it has been shown that the diversity of filamentous bacteria is much higher in industrial wastewater treatment plants (Snaidr *et al.*, 2002, Eikelboom and Geurkink, 2001, Eikelboom and Geurkink, 2002, Levantesi *et al.*, 2004).

Table 2.2 reports the main characteristics and kinetic properties of the filamentous microorganisms that have been studied in the last than twenty years. The following points can be highlighted.

- All the isolated filaments are aerobic and are not able to denitrify nitrate completely to nitrogen gas (some can reduce nitrate partially to nitrite).
- None of the filamentous bacteria are reported to be fermentative, except for *Nostocoida limicola* I.
- Many filamentous bacteria are heterotrophs and utilise mainly simple organic substrates such as carbohydrates, lipids, and proteins. Some species are characterised by high nutritional versatility, such as *Beggiatoa* and *Thiothrix,* which are able to grow heterotrophically, mixotrophically (simultaneous growth on organic carbon and sulphur reduced compound), and autotrophically (growth only on sulphur reduced compounds as energy source).
- The maximum growth rates of filamentous bacteria are low (typically in the range 0.4 - 3.0 d^{-1}), especially if compared with growth rates characteristic of floc-forming bacteria (for instance *Acinetobacter* has a maximum growth rate at 20 °C on acetate 13.5 d^{-1}, Authors data).
- All the examined bacteria store at least one intracellular polymer, such as lipid inclusions (i.e. PHA) and polyphosphates. Some organisms contain sulphur granules and others (*H. hydrossis)* store polysaccharides.
- Molecular probes have already been defined (Table 2.3) for some filamentous bacteria involved is bulking and other important AS bacteria as well; these probes should be used to complement the more traditional morphological information (Table. 2.4).

Current knowledge of the kinetic properties and metabolic versatility of filamentous bacteria can not be considered satisfactory, mainly because only a few isolates have been studied, and often using simple substrates. This incomplete picture is primarily a results of the difficulties involved in isolation and conducting experiments with these organisms. It is difficult to obtain isolates, and often, they slow down and eventually stop growing in pure culture. In the following paragraphs more recent information on two interesting bacteria, *M. parvicella* and *Thiothrix* sp., the only two filaments that have been phylogenetically and physiologically characterised, is presented. Information are also provided on the *Nostocoida limicola* group, whose representatives have been recently isolated, even though only partially from a physiological point of view.

Table 2.2. Kinetic and physiological properties of filamentous bacteria isolates.

Bacterium	Aerobic respiration	Anaerobiosis	Substrates	μ max (d^{-1})	Ks (mg/L)	Storage products	Reference
Sphaerotilus natans	C	$NO_3 \rightarrow NO_2$	AL, AC, CA	6.5$^{(1)}$	10.0	PHB	Richard *et al.*, 1985 William and Unz, 1985
Type 1701	C	some strains: $NO_3 \rightarrow NO_2$	AL, AC, CA	2.6$^{(1)}$	2.0	PHB	Richard *et al.*, 1985 Williams and Unz, 1985
Haliscomenobacter hydrossis	C	-	GLU, LAT	1.2-2.2$^{(1)}$	5.0	PS	Krull, 1977
Beggiatoa	C, A, M	-	AC	n.d.	n.d.	PHB, PP, S	Williams and Unz, 1985
Thiothrix spp.	M	-	AC	n.d.	n.d.	PHB, PP, S	Williams and Unz, 1985
Thiothrix, strain CT3	M	$NO_3 \rightarrow NO_2$	AC + HS^-	2.6$^{(2+3)}$	n.d.	PP, S	Tandoi *et al.*, 1994
Thiothrix, strain CT3	C	$NO_3 \rightarrow NO_2$	AC	2.5$^{(2)}$	n.d.	PHB	Tandoi *et al.*, 1994
Thiothrix, strain CT3	A	$NO_3 \rightarrow NO_2$	HS^-	1.8$^{(3)}$	n.d.	S	Tandoi *et al.*, 1994
Type 021N spp.	C, M	$NO_3 \rightarrow NO_2$	CA, AC	3.75$^{(2)}$	0.07$^{(4)}$	PHB, PP, S	Williams and Unz, 1985 Van Niekerk *et al.*, 1987
Gordona amarae	C	-	CA, AC	2.5$^{(2)}$	0.5$^{(4)}$	PHB, PP	Blackall *et al.*, 1991
M. parvicella	C	-	AGL	1.4*$^{(4)}$ 0.6**$^{(4)}$	n.d.	LP	Slijkhuis and Deinema, 1988
M. parvicella, strain RN1	C	$NO_3 \rightarrow NO_2$	AGL, OL	0.6$^{(5)}$	n.d.	LP, PP	Tandoi *et al.*, 1998a
M. parvicella, strains RN1 and 4B	C	$NO_3 \rightarrow NO_2$	AC+mixture $^{(5)}$	-	n.d.	PHAs (in anaerobic and anoxic conditions)	Rossetti *et al.*, 2002
N.limicola spp.	C	$NO_3 \rightarrow NO_2$	CA, PEPT	n.d.	n.d.	PHB	Blackall *et al.*, 1999
N.limicola I	C	F					Seviour *et al.* 2001
N.limicola II	C	-					Snaidr *et al.*, 2002

(§) Nitrate as electron acceptor (final product is NO_2^-, none of the filamentous bacteria is able to reduce nitrate to nitrogen gas); *estimated in continuous reactor, **estimated in batch reactor; n.d. not determined; (1) glucose; (2) acetate; (3) HS^-; (4) oleic acid; (5) mixture of substrates constituted by glucose, yeast extract, proteose peptone, pyruvic acid, starch and casaminoacids; C: Chemoorganotroph, A: Autotrophs, M: Mixotrophs; F: Fermentative, AL: Alcohol, AC: organic acids, CA: carbohydrates, GLU: glucose, LAT: lactose, AGL: long chain fatty acids; OL: oleate; PEPT: peptone; PS: polysaccharides; PP: polyphosphates; LP: lipids.

Table 2.3. Main molecular probes available for in situ identification of bacteria present in wastewater treatment systems.

	Microrganism	Probe	Probe sequence (5' - 3')	Target and position*	% Formammide	Reference
Filamentous bacteria	*M.parvicella*	MP223	GCCGCGAGACCCTCCTAG	16S, 223-240	20	Erhart *et al.*,1997
	Thiothrix nivea	TN1	CTCCTCTCCCACATTCTA	16S, 652-669	45	Wagner *et al.*, 1994a
	H. hydrossis	HHY	GCCTACCTCAACCTGATT	16S, 655-672	20	Wagner *et al.*, 1994a
	Type 021N	21N	TCCCTCTCCCAAATTCTA	16S,652-669	35	Wagner *et al.*, 1994a
	S.natans	SNA	CATCCCCCTCTACCGTAC	16S,656-673	45	Wagner *et al.*, 1994a
	Acinetobacter sp.	ACA	ATCCTCTCCCATACTCTA	16S, 652-669	35	Wagner *et al.*, 1994b
	N.limicola I	LIMI91	CGCCACTATCTTCTCAGT	16S, 91-108	20	Liu and Seviour, 2001
	N.limicola II	NLIMII175	GGCTCCGTCTCGTATCCG	16S, 175-192	40	Liu and Seviour, 2001
		NLIMII192	AGACTTTCCAGACAGGAG	16S, 192-209	20	
	N.limicola in industrial systems:					
	Candidatus "Alysiosphaera europaea"	Noli-644	TCCGGTCTCCAGCCACA	16S, 644-660	35	Snaidr *et al.*, 2002
	N.limicola III	NLIMIII301	CCCAGTGTGCCGGGCCAC	16S, 301-318	20	Liu and Seviour, 2001
		NLIMIII729	AGCATCCAGAACCTCGCT	16S, 729-746	20	
		NLIMIII830	CCATCGGCGAGCCCCCTA	16S, 830-847	20	
Nitrifiers	Ammonia oxidisers:					
	Nitrosospira and Nitrosomonas	NSO190	CGATCCCCTGCTTTTCTCC	16S, 190-208	55	Schramm *et al.*, 1998
	Nitrite oxidisers:					
	Nitrospira	NSR1156	CCCGTTCTCCTGGGCAGT	16S, 1156-1173	30	Schramm *et al.*, 1998
	Nitrobacter spp.	NIT2	CGGGTTAGCGCACCGCCT	16S, 1433-1450	40	Wagner *et al.*, 1996
	Nitrobacter spp.	NIT3	CCTGTGCTCCATGCTCCG	16S, 1030-1047	40	Wagner *et al.*, 1996
Denitrifiers (some)	*Pseudomonas* spp	Ps	GCTGGCCTAGCCTTC	23S, 1432-1446	35	Schleifer *et al.*, 1992
	Paracoccus spp.	PAR651	ACCTCTCTCGAACTCCAG	16S, 651-668	40	Neef *et al.*, 1996
	Paracoccus spp.	PAR1244	GGATTAACCCACTGTCACC	16S, 1244-1283	20	Neef *et al.*, 1996
	Paracoccus spp.	PAR1457	CTACCGTGGTCCGCTGCC	16S, 1457-1474	35	Neef *et al.*, 1996
	Paracoccus denitrificans	Pdv198	CTAATCCTTTGGCGATAAATC	16S, 198-232	20	Neef *et al.*, 1996
	Paracoccus denitrificans	Pdv1031	CCTGTCTCCAGGTCACCG	16S, 1031-1048	35	Neef *et al.*, 1996

*according enumeration of *E. coli* in Brosius *et al.*, 1978.

Table 2.4: main useful characteristics for the identification of the most common "traditional" filamentous bacteria causing bulking and foaming problems.

	Phylogenetic affiliation	Gram Staining	Neisser Staining	Ø µm	Sheath presence	Sulphide oxidation Test	Cell shape	Key morphological peculiarity
S natans	Betaproteobacteria	-	-	1.0-1.4	+	-	round	False branching
Type 1701		-	-	0.6-0.8	+	-	round e	Attached growth
Type 0041		V*	-/+	1.4-1.6	+	-	square	Gram Staining
Type 021 N (now T eikelboomii)	Gammaproteobacteria	-	-	1.0-2.0	-	+	barrel	cell shape
Thiothrix	Gammaproteobacteria	-	-	1 -2.5	+	+	square	Sulphur granules presence/production
Type 1851	CF	V	-	0.8	+	-	indistinguishable	Filaments bundle assembled
Type 0092	Not yet certain isolates	-	+	0.8-1.0	+	-	indistinguishable	Neisser Staining
M. parvicella	Actinobacteria, in High G+C Gram positive Bacteria	+	-	0.8	Non ascertained	-	indistinguishable	Gram Staining
Nocardioforms	Several genera and species belonging to different divisions of Bacteria	+	-	1.0	-	-	indistinguishable	True Branching, Gram Staining
N limicola I	Low G+C Gram positive Bacteria	+	-					Gram Staining
N limicola (II) (several genera and species)	High G+C Gram positive Bacteria and Alfaproteobacteria	-/+	-/+	1.2-1.4	-	-	discoid	Cell Shape
H hydrossis	CF	-	-	0.5	+	-	indistinguishable	Filaments dimension and typical bent

* V: variable, parts of the cells are positive, parts are negative; -/+: it can be positive or negative

2.5 MICROTHRIX PARVICELLA

M. parvicella is the most common cause of bulking and foaming in AS system. Since its first detailed description (van Veen, 1973) many attempts have been made to characterize it and control its proliferation. Nevertheless, only a few pure cultures studies have been performed to date. This lack of data is mainly due to difficulties with its isolation, cultivation, and its characteristic slow growth. Information on the in situ uptake of substrates has been determined by MAR (microautoradiography). From this information and numerous pilot and full-scale experiments, several hypothesis have been formulated to explain its proliferation in AS systems and consequently, to define methods to control its growth. In particular, as far as the physiology of the organism is concerned, the RN1 strain is the only isolate for which the taxonomic position and basal physiological and kinetic characterisation have been described (Tandoi *et al.*, 1998; Rossetti *et al.*, 2002).

Identification of M. parvicella

M. parvicella is a thin (0.8-1.0 μm long), unsheathed Gram positive bacterium with an irregular and winding appearance. Normally this organism does not favor attached growth and it is never branched. *M. parvicella* often accumulates polyphosphate and can also accumulate PHAs. Due to its distinctive morphology and particularly, to its characteristic response to Gram staining, the presence of this filamentous bacterium in AS is easy to detect. The few isolates in existence have very similar 16S rDNA sequences (99.6 %). Molecular probes are available and allow precise identification of this filamentous bacterium in sludge samples.

Table 2.5. Summary of the growth and storage capabilities of pure and mixed cultures of *M. parvicella* under different metabolic conditions.

	GROWTH		STORAGE		REFERENCE*
	Mixed culture	Pure culture	Mixed culture	Pure culture	
AEROBIOSIS	+	+	PHA, PP (1, 2, 3)	PHA PP (3)	(1) Tandoi *et al.*, 1998
ANAEROBIOSIS	- (3)	- (3)	PHA (2, 3)	PHA (3)	(2) Andreasen *et al.*, 2000, (3) Rossetti *et al.*, 2002
ANOXIC CONDITIONS					
Nitrate	- (3)	- (3)	PHA (3)	PHA (3)	
Nitrite	- (3)	- (3)	PHA (3)	PHA (3)	

PHA: poly-hydroxyalkanoates; PP: polyphosphates

Physiology of M. parvicella

The first isolate described in detail was isolated by Slijkhuis and coworkers (1982). Unfortunately, this isolate lost its viability before the sequence of its 16S rDNA gene was determined. Two new strains of *M. parvicella* were isolated approximately 10 years later (Blackall *et al.*, 1994, Rossetti *et al.*, 1997, Rossetti *et al.*, 2002). The limited number of strains, were sufficient to describe its basic physiology and kinetic properties, and to give to the organism name the status of "Candidatus" (Blackall et al., 1996), a first step in naming organisms when only few strains and information are available. Table 2.5 reports the main properties of this organism obtained in pure and mixed culture studies. It is possible to conclude that one of the main factors promoting its growth is its ability to benefit from the presence of anoxic and anaerobic zones, where it can store organic carbon. This unusual storage phenomenon is presenting a main trend area of research. For a recent review see Rossetti et al., 2005.

2.6 THIOTHRIX

Thiothrix is a bacterium that can grow mixotrophically, heterotrophically or litotrophically. This filamentous bacterium is often present in AS plants treating sewage characterized by a high concentration of reduced sulphur compound, such as sulphides and thiosulphates. These filamentous bacteria can cause bulking under such conditions (Jenkins *et al.*, 1993). Bacteria belonging to the genus *Thiothrix* are characterized by distinctive morphological features, such as growth in unsheathed filaments, accumulation of sulphur granules when incubated in the presence of reduced sulphur compounds, organization in rosette-like arrays, and production of gliding gonidia from the filament apex. The filamentous bacteria of the genus *Thiothrix* were first described by Winogradsky in 1888 (cited in Odintsova et al., 1993). He hypothesised that *Thiothrix* might be autotrophic, but the neotype strain, *Thiothrix nivea,* was thought to be an obligate mixotroph, requiring both a reduced sulphur compound and an organic substrate for growth. All of the other strains belonging to this genus have been reported to be mixotrophic or heterotrophic (Larkin and Shinabarger, 1983). Recently, it has been shown that an isolate from AS, strain CT3, and another strain, strain I (William & Unz, 1985) are also capable of growing autotrophically with reduced sulphur compounds as the sole energy source (Rossetti *et al.*, 2003). Another chemolithotrophic strain of *Thiothrix* was obtained from a hydrogen sulphide spring and phylogenetically identified as *T. ramosa* (Polz, 1996). The type species of the genus is now *T. nivea,* the autotrophic *T. ramosa* name is still invalid, but four new species were proposed. They are *T. fructosivorans* (for the Q , I and CT3), *T. unzii* (for strain A1), *T. eikelboomii* (for the Eikelboom Type 021N strains, type strain AP3), and *T. defluvii* (for an Australian *Thiothrix* isolate called Ben57). All of new species are from activated sludge sewage treatment plants. It is necessary to underline that the important morphotype, known as Type O21N and commonly found in activated sludge samples, has been shown to belong to the same *Thiothrix* genus (Howarth *et al.*, 1999), and it is now named *Thiothrix eikelboomii.* Significative differences between *Thiothrix* strains recalling Type 021 N and *Thiothrix nivea*-like strains were found: the first were able to utilise several fatty acids and many carbohydrates while the second mainly fatty acids (Aruga et al. 2002).

2.7 NOSTOCOIDA LIMICOLA

Since the first description of "*Nostocoida limicola*" (Morphotypes I, II and III differing one each other for dimensions and Gram and Neisser staining) in domestic WWTPs, sequences of the 16S rRNA gene from several isolates showed that this group contained phylogenetically unrelated bacteria, including members of the Low G+C mol% Gram positive bacteria, the *Actinobacteria* and the *Planctomycetales* (Blackall *et al.*, 2000, Seviour *et al.*, 2002). On the other hand, the Gram negative *N. limicola* seen in industrial treatment plants belong to the *Alphaproteobacteria* (Eikelboom & Geurkink, 2002, Levantesi *et al.*, 2004), and these have only rarely been reported in domestic treatment plants (Seviour *et al.*, 2002).

Very few physiological data are available at present on the *Nostocoida* isolates: many molecular probes are today available to identify them *in situ*, but precise correlations between these filamentous bacteria and the conditions favouring their growth, are still under investigation.

2.8 POLYPHOSPHATE ACCUMULATING ORGANISMS (PAO)

Over the past decades, design advances have taken place, resulting in plants that remove not only carbonaceous compounds and nitrogen, but also phosphorus. Microorganisms with the ability to take up and store polyphosphate are exploited for removing phosphorus from wastewater in a process known as enhanced biological phosphorus removal (EBPR). Polyphosphate accumulating organisms (PAOs) are selectively enriched under alternating anaerobic/aerobic conditions. Wastewater containing phosphorus and soluble organic carbon compounds is mixed with sludge under anaerobic conditions. The mixture is maintained under anaerobic conditions for a period of time, then passed to an aerobic zone before going to the secondary sedimentation tank for sludge settlement and recycling. At present, neither a detailed understanding of this process nor a pure culture of PAOs is available. Nevertheless, a biochemical model has been postulated that describes EBPR (Mino *et al.*, 1987; Pereira *et al.*, 1996). According to this model, the PAOs have a selective advantage over other aerobic heterotrophic bacteria in the anaerobic-aerobic sequence of the EBPR system. In the anaerobic zone, PAOs utilise part of their stored polyphosphate and glycogen as sources of energy and reducing power for the uptake of soluble organic carbon (typically in the form of volatile fatty acids, VFAs) and for the synthesis and storage of poly-β-hydroxyalkanoates (PHAs). In the aerobic zone, these same cells take up phosphorus (both that which they released and that from the influent) and store it as polyphosphate. PAOs also utilise anaerobically stored PHA as a carbon source for cell growth and for replenishing glycogen.

Based on cultivation experiments, the genus *Acinetobacter* was traditionally believed to be the only representative of PAO in activated sludge. Today, however, it has been proven that the presence of this genus has been overestimated. In addition, to date only one PAO has been unequivocally identified, and it belongs to the beta 2-subclass of Proteobacteria (*Candidatus Accumulibacter phosphatis*) (Hesselmann *et al.*, 1999; Crocetti *et al.*, 2000). Phosphorus and carbon transformations indicative of EBPR have been proven to occur in these bacteria when they are in mixed EBPR sludge cultures (Crocetti et al., 2000). For recent reviews see: Blackall *et al.*, 2002 and Seviour *et al.*, 2003.

2.9 GLYCOGEN ACCUMULATING ORGANISMS (GAO)

Sometimes for unknown reasons, EBPR fails and the plant effluent contains high levels of phosphorus. Failures often occur despite seemingly correct conditions and typically in periods following stable, efficient EBPR. A potential reason for EBPR failure is the presence of microorganisms that use stored compounds, such as glycogen, to compete with the PAOs for carbon under anaerobic conditions. Although these microorganisms have the capacity to carry out the anaerobic and aerobic carbon transformations consistent with those of PAOs, they cannot cycle phosphorus. Because these organisms produce glycogen in the aerobic zone and apparently obtain their energy from glycogen, they are known as glycogen accumulating organisms (GAOs) (Satoh *et al.*, 1994).

The GAO phenotype has not been observed in pure culture, but it has been demonstrated by chemical analysis in highly enriched cultures (Fukase *et al.*, 1985; Satoh *et al.*, 1994; Bond *et al.*, 1999b) and by microscopy at a cellular level (Crocetti *et al.*, 2002). Tetrad -forming organisms (TFOs), previously known as "G bacteria", were implicated as GAOs in EBPR by Cech and Hartman (1993). Recent molecular and physiological analyses of isolates with TFO morphology, however, revealed high phylogenetic diversity and provided no support for their role as GAOs (Falvo *et al.*, 2000;

Seviour *et al.*, 2000). To date, one GAO has been unequivocally identified, and it belongs to a novel cluster of bacteria in the *gammaproteobacteria*, *Competibacter phosphatis* (Nielsen *et al.*, 1999; Dabert *et al.*, 2001; Crocetti *et al.*, 2002). For recent reviews see: Blackall *et al.*, 2002; Seviour *et al.*, 2003.

2.10 NITRIFIERS

Since most of the nitrogen in wastewater influent is present either as urea (easily hydrolysed to ammonia) or as ammonia, nitrifying bacteria play an important role in nitrogen removal in wastewater treatment plants.

Nitrifiers are extremely slow-growing microorganisms and are difficult to cultivate. Because of their sensitivity to disturbances in pH and temperature, breakdown of the nitrification process occurs frequently.

The genus *Nitrosomonas* has long been considered as the model ammonia-oxidizer. Other genera of nitrifying bacteria have been detected in nitrifying activated sludge and biofilms by molecular analysis and include *Nitrosococcus*, *Nitrosospira* and *Nitrosovibrio*. Likewise traditionally members of the genus *Nitrobacter* were considered to be the most important nitrite-oxidising bacteria in wastewater treatment plants. Surprisingly, this genus could not be detected in activated sludge at different nitrifying wastewater treatment plants when monitored by FISH with specific 16S rRNA-targeted probes. Using a full-cycle rRNA approach, the occurrence of novel, yet uncultured *Nitrospira-* like nitrite-oxidising bacteria have been found (Juretschko *et al.*, 1998; Daims *et al.*, 2001). Different survival strategies have been postulated for explaining the predominance of *Nitrospira*-like bacteria over *Nitrobacter* in most wastewater treatment plants. *Nitrospira*-like nitrite-oxidisers are characterised by a low μ max and are, according to results obtained by Schramm *et al.* (1999), k-strategist. They are thus well-adapted to low nitrite and oxygen concentrations. *Nitrobacter* was postulated to be a relatively fast-growing r-strategist with low affinities for nitrite and oxygen. Since nitrite concentrations are often low in most aeration tanks, *Nitrospira* –like organisms should be able to successfully outcompete *Nitrobacter* species in these systems.

2.11 DENITRIFIERS

The ability denitrify is common among different groups of bacteria. Nevertheless, it is certain that filamentous bacteria are not denitrifiers. Because many floc-formers are able to denitrify the use of specific molecular probes to detect these organisms is difficult unless probes are designed for only specific denitrifying populations (Neef *et al.*, 1996, see Table 2.3 and Ginige *et al.,* 2004).

2.12 QUANTITATIVE ESTIMATION OF FILAMENTOUS BACTERIA IN ACTIVATED SLUDGE

While molecular-based techniques are continuing to develop, quick and simple estimation of filament numbers can still be obtained by optical microscopy. Pitt and Jenkins (1990) proposed a simple and rapid method for estimating the presence of *Nocardia*-like organisms in activated sludge. This technique is suitable for both foams and mixed liquor samples. The technique to estimate Nocardioform is described below, and it can also be applied for other filaments.

Nocardioform estimation is possible by microscopic analysis, and represents an easy tool to evaluate the effectiveness of specific actions in plant.

A) Place 80 µl of mixed liquor or diluted foam on several microscopic slides whose edges have been marked with a glass marker at three points along their length (Figure 2.3).
B) Spread the sample evenly over the slides and allow it to air-dry. Examine under phase contrast at 100X magnification to check the even distribution of the sample on the slide. Five slides per sample are needed.
C) Gram stain the smears and then count the five slides under oil immersion and bright field microscopy using an eyepiece graticule with a ruled line

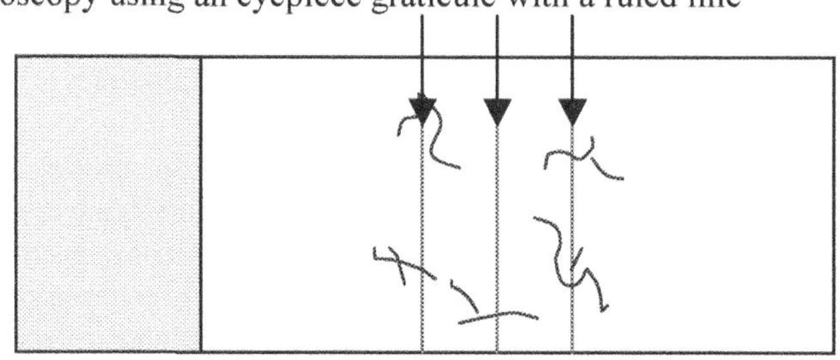

Figure 2.3 Schematic marks at the three points along the slide length.

1) Locate the mark on the edge of the slides and line up the eyepiece line with it. Count any intersection made by Gram-positive branched filaments.
2) Move across the slide to the opposite edge counting all the intersections of Gram positive filaments.
3) Repeat the last two steps at the other two marks present on the slides and calculate the total intersections in the three lines: Average the count for the five slides.
4) Knowing the MLSS concentration of the sample, the Nocardioform count will be:

$$\text{n}^\circ \text{ of intersections/g VSS} = \frac{\text{Average of intersections x 1000 x1000}}{\text{80 x MLVSS (g/l)}}$$

Normally, foaming problems start when this number is above 10^5 intersections/g VSS.

2.13 FISH ANALYSIS

In this paragraph, we provide the protocol for FISH analysis (Amann, 1995) for application in wastewater microbiology. During the last years, many rRNA targeted oligonucleotide probes for *in situ* detection of bacteria present in biological treatment systems, have been designed. Some of these probe sequences and stringency conditions can be easily accessed on-line using probeBase (http://www.microbial-ecology.net/probebase, Loy *et al.*, 2003). By applying these probes under the recommended hybridisation conditions, the respective target microorganisms can be specifically identified in appropriately-fixed activated sludge samples within about three hours. The technique consists basically of three main steps: fixation and dehydration of the sample, hybridisation with the probes and successive washing, and finally, sample examination by epifluorescent microscopy. Sample fixation assures rRNA content preservation and preservation of cellular integrity even during exposure to high temperatures and high osmotic gradients during the hybridisation step. The standard fixation procedure for both Gram positive and Gram negative bacteria is described below. The paraformaldehyde fixation is most suitable for Gram-negative cells. For samples containing Gram-positive cells, also fix sample aliquots with ethanol only. It is wise to use both fixations methods with unknown samples.

Sample fixation

Paraformaldehyde fixation (Gram negative cells):

Solutions: 3 x and 1x Phosphate Buffered Saline (PBS, it can be either bought or made according Amann et al., 1995) and 4% paraformaldehyde

Preparation of 4% paraformaldehyde:

- heat 33 ml dist. H_2O to 60-65°C
- add 2 g paraformaldehyde while stirring
- add 10 N NaOH drop-wise until paraformaldehyde is dissolved completely
- add 16.5 ml 3x PBS buffer
- adjust the pH to 7.2-7.4 at 20°C
- filter through 0.45 µm filter
- use within 24 h, store on ice until use

Fixation:

- harvest cells by centrifugation
- wash the pellet with 1x PBS, centrifuge again and resuspend pellet in an appropriate volume of fresh 1x PBS
- add 3 volumes of fresh paraformaldehyde solution
- incubate for 3 h or over night at 4°C
- spin down the cells (5 min)
- discard supernatant
- wash the pellet with 1x PBS, centrifuge again
- resuspend the pellet in fresh 1x PBS
- add 1 volume of ice-cold $EtOH_{abs.}$
- store at –20°C

Ethanol fixation (Gram positive cells):

- harvest cells by centrifugation
- wash the pellet with 1x PBS, centrifuge again
- resuspend the pellet in an appropriate volume of fresh 1x PBS
- add 1 volume of ice-cold $EtOH_{abs.}$
- store at –20°C

Quick sample fixation

While the previous protocol is recommended for long-term storage of samples, a rapid and simpler procedure can be followed instead to fix activated sludge samples for more rapid screening by FISH analysis (the fixed sample can be suitable for use with FISH within some months). With this quick method, resuspention of the sample in PBS is avoided and the fixatives are added directly to the sludge sample. Moreover, instead of using paraformaldehyde as fixative for Gram positive cells, 37% formaldehyde is utilized. The procedures are given below; as above, it is suggested to fix the sludge sample using both procedures.

Formaldehyde (Gram negative cells):
Add 5 ml of 37% formaldehyde to 45 ml of activated sludge. Add 50 ml of 96% ice-cold ethanol. Store the fixed sample at – 20°C.

Ethanol fixation (Gram positive cells):
Add 55 ml of 96% ice-cold ethanol to 45 ml of activated sludge sample. Store the fixed sample at – 20°C.

Standard protocol for hybridization on slides

For FISH analysis, slides in which a hydrophobic coating (Teflon) separates 10-12 glass surface windows are required (i.e. Teflon coated slides from Marienfeld, Germany). This coating permits different hybridizations to be performed on the same slide by preventing the mixing of probes applied to adjacent spots on the slide (Figure 2.4a).

- prepare specimen on a teflon coated slide: spot fixed cells (2-15 µL, amount depending on the cell concentration) and dry for about 10 min at 46°C
- in the meantime, prepare the hybridization buffer (see below) and keep at room temperature
- dehydrate specimen on the slide in an increasing ethanol series (3 min each in 50, 80 and 100 % ethanol)
- in the meantime, thaw the oligonucleotide probes
- drop 10 µL of hybridization buffer onto the wells
- add 0.5 µL of each probe (working solution, 50 ng/µL) without scratching the teflon surface. Store frozen probe aliquots at -20°C.
- prepare a hybridization tube (50 ml Falcon tube) by folding a piece of paper, putting it into the tube and pouring the remaining hybridization buffer onto the paper (Figure 2.4b)
- immediately transfer the slide into the hybridization tube and incubate in the hybridization oven (46 ± 0.5 °C) for 1.5 hours
- in the meantime, prepare the washing buffer (see below) and preheat this buffer to 48°C (in a water bath)
- rinse the hybridization buffer from the slide with the washing buffer and incubate the slide in the washing buffer for 10 min in a 48 °C water bath
- remove the washing buffer with distilled water and dry the slide quickly using compressed air.

FISH ANALYSIS

- **Sample fixation** (ethanol and paraformaldehyde fixation) to keep cell integrity
- **Hybridization**
  - add 3-5 ml of fixed sample to teflon coated slide and let it dry

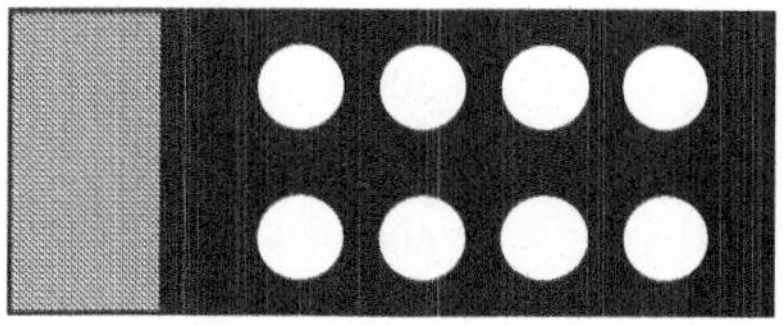

  - dehydrate specimen on the slide in an increasing ethanol series (3 min each in 50, 80 and 100 % ethanol)
  - add 10 ml hybridization buffer and 0.5 ml molecular probe (50 ng/ml)
  - incubation for 1-2 hours at 46°C keeping the slide horizontally in a 50 ml screw-cap tube containing tissue paper moistened with hybridization buffer

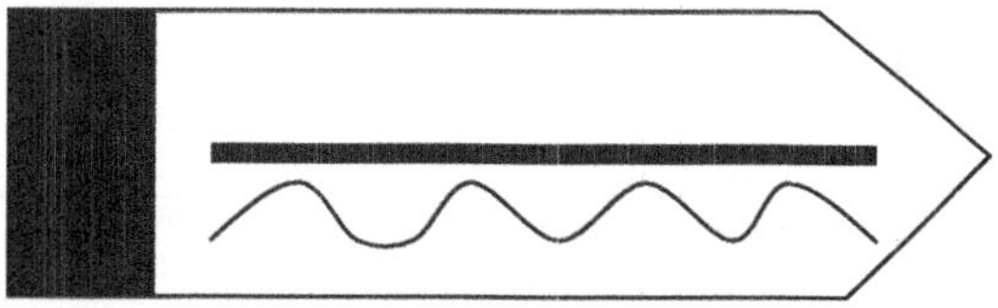

- **Washing** at 48°C
- **Epifluorescence microscopic observation**

Figure 2.4 FISH Analysis.

Hybridization buffer composition

Prepare 2 ml of hybridization buffer for each slide. Add to 2 ml Eppendorf reaction tube:

- 5 M NaCl 360 µL
- 1 M Tris/HCl pH 8,0 40 µL
- add formamide and ultra pure water, depending on the applied stringency according the following table:

% Formamide (v/v)	Formamide [µL]	Water [µL]
0	0	1.600
5	100	1.500
10	200	1.400
15	300	1.300
20	400	1.200
25	500	1.100
30	600	1.000
35	700	900
40	800	800
45	900	700
50	1.000	600
60	1.100	500
65	1.200	400
70	1.300	300

- 10 % (w/v) SDS 2 µL

Washing buffer composition

Mix in a 50 ml Falcon tube and preheat to 48°C prior to use:

- 1 ml 1 M Tris/HCl pH 8,0
- 5 M NaCl according to the following table:

% Formamide in hybridization buffer	[NaCl] in mol/L	NaCl [µL]
0	0,900	9.000
5	0,636	6.300
10	0,450	4.500
15	0,318	3.180
20	0,225	2.150
25	0,159	1.490
30	0,112	1.020
35	0,080	700
40	0,056	460
45	0,040	300
50	0,028	180
55	0,020	100
60	0,008	40

- 50 µL 10 % (w/v) SDS
- 500 µL 0.5 M EDTA only if hybridisation buffer contains 20% or more formamide
- fill to the 50 mL mark with water

Sample examination by epifluorescent microscopy

Mount slides by adding a few drops of Vectashield antifading; put a coverslip on the slide and wait until the antifading has spread over the slide. View the sample with an epifluorescence microscope equipped with suitable filter sets e.g., Zeiss filter sets n. 10 for Fluorescein, n. 15 for CY3 and n. 1 for DAPI staining. Slides embedded in antifading can be stored at -20 °C for a long period of time (months).

2.14 REFERENCES

Amann R.I. (1995). In situ identification of micro-organisms by whole cell hybridization with rRNA-targeted nucleic acid probes. *Molecular Microbial Ecology Manual,* 3.3.6 1-15.

Blackall L.L., Tandoi V., Jenkins D. (1991). Continuous culture studies with Nocardia amarae from activated sludge and their implications for *Nocardia* foaming control. *Res J WPCF*, **63**, 44-50.

Aruga S., Kamagata Y., Kohno T., Hanada S., Nakamura K., Kanagawa T. (2002). Characterization of filamentous Eikelboom Type 021N bacteria and description of *Thiothrix disciformis* sp. nov. and *Thiothrix flexilis* sp., IJSEM, **52**, 1309–1316

Blackall L.L., Seviour E.M., Bradford D., Rossetti S., Tandoi V., Seviour R.J. (2000). Candidatus *Nostocoida limicola* II a filamentous bacterium from activated sludge. *IJSEM*, **50**, 703-709.

Blackall L.L., Seviour E.M., Cunningham M.A., Seviour R.J., Hugenholtz P. (1994). *Microthrix parvicella* is a novel deep branching member of the actinomycetes subphylum. *Syst Appl Microbiol,* **17**, 513-518.

Blackall L.L., Stratton H., Bradford D., Sjörup C., Del Dot T., Seviour E.M., Seviour R.J. (1996). Candidatus *Microthrix parvicella* - a filamentous bacterium from activated sludge sewage treatment plants. *Intern J System Bacteriol*, **4**, 344-346.

Blackall L.L., Crocetti G., Saunders A. M., Bond P.L. (2002). A review and update of the microbiology of enhanced biological phosphorus removal in wastewater treatment plants. Antonie van Leeuwenhoek, **81**, 681-691.

Bond P.L., Erhart R., Wagner M., Keller J., Blackall L.L. (1999). Identification of some of the major groups of bacteria in efficient and non efficient biological phosphorus removal activated sludge systems. *App Environ Microbiol*, **65**, 4077-4084.

Brosius J., Palmer M.L., Kennedy P.J., Noller H.F. (1978). Complete nucleotide sequence of a 16S ribosomal RNA gene from *Escherichia coli. Proc Natl Acad Sci* USA. 75. 4801-4805.

Cech J.S., Hartman P. (1993). Competition between polyphosphate and polysaccharide accumulating bacteria in enhanced biological phosphorus removal systems. *Water Res*, **27**, 1219-1225.

Crocetti G R., Hugenholtz P., Bond P. L., Schuler A., Keller J., Jenkins D., Blackall L.L. (2000). Identification of polyphosphate-accumulating organisms and design of 16S rRNA-directed probes for their detection and quantification. *Appl Environm Microbiol,* **66**, 1175-1182.

Crocetti G. R., Banfield J. F., Keller J., Bond P. L., Blackall L. L. (2002). Glycogen-accumulating organisms in laboratory-scale and full-scale wastewater treatment processes. *Microbiology*, **148**, 3353-3364.

Dabert P., Sialve B., Delgenès J.P., Moletta R., Godon J.J. (2001). Characterization of the microbial 16S rDNA diversity of an aerobic phosphorus-removal ecosystem and monitoring of its transition to nitrate respiration. *App Microbiol Biotech*, **55**, 500-509.

Daims H., Nielsen J.L., Nielsen P.H., Schleifer K.H., Wagner M. (2001). *In situ* characterization of *Nitrospira*-like nitrite-oxidising bacteria active in wastewater treatments plants. *Appl Environm Microbiol*, **67**, 5273-5284.

Eikelboom D.H. (1975). Filamentous organisms observed in activated sludge *Water Res*, **9**, 365-388.

Eikelboom D.H., van Buijsen H.J.J. (1981). *Microscopic Sludge Investigation Manual,* 1st edn. The Netherlands: TNO Research Institute for Environmental Hygiene.

Eikelboom D.H., van Buijsen H.J.J. (1983). *Microscopic Sludge Investigation Manual* 2nd edn. The Netherlands TNO Research Institute for Environmental Hygiene.

Eikelboom D.H. (2000). Process Control of Activated Sludge Plants by Microscopic Investigations. IWA Publishing, London, UK.

Eikelboom D.H., Geurkink B. (2001). Catalogue of filamentous micro-organisms occurring in industrial activated sludge plants. TNO-MEP, Report R 2001/090.

Eikelboom D.H., Geurkink B. (2002). Filamentous micro-organisms observed in industrial activated sludge plants *Water Sci & Technol* **46**(1-2), 535-542.

Erhart R., Bradford D., Seviour R.J., Amann R., Blackall L.L. (1997). Development and use of fluorescent in situ hybridization probes for the detection and identification of *Microthrix parvicella* in activated sludge. *Syst Appl Microbiol,* **20,** 310-318.

Falvo A., Levantesi C., Rossetti S., Seviour R.J., Tandoi V. (2001). Synthesis of intracellular storage polymers by *Amaricoccus kaplicensis* a tetrad forming bacterium present in activated sludge. *J Appl Microbiol*, **91**, 299-305.

Fukase T., Shibata M., Miyaji Y., (1985). The role of an anaerobic stage on biological phosphorus removal. *Water Sci & Technol*, **17**(2), 69-80.

Ginige M.P., Hugenholtz P., Daims H., Wagner M., Keller J., Blackall L.L. (2004). Use of stable-isotope probing full-cycle rRNA analysis fluorescent in situ hybridisation-microautoradiography to study a methanol-fed denitrifying microbial community. *Appl Environ Microbiol*, **70**, 588-596.

Hesselmann R.P.X., Werlen C., Hahn D., van der Meer J.R, Zehnder A.J.B. (1999). Enrichment phylogenetic analysis and detection of a bacterium that performs enhanced biological phosphate removal in activated sludge. *Syst Appl Microbiol*, **22**, 454-465

Howarth R., Unz R.F., Seviour E.M., Seviour R.J., Blackall L.L., Pickup R.W., Gwyn Jones J., Yaguchi J., Head I.M. (1999). Phylogenetic relationships of filamentous sulfur bacteria (*Thiothrix* spp. and Eikelboom Type 021N bacteria) isolated from wastewater-treatment plants and description of *Thiothrix eikelboomii* sp nov. *Thiothrix unzii* sp nov. *Thiothrix fructosivorans sp* nov and *Thiothrix defluvii sp* nov. *Int J Syst Bacteriol*, **49**, 1817-1827.

Jenkins D., Richard M.G., Daigger G.T., (1984). Manual on the causes and control of activated sludge bulking and foaming. 1st edn. Lewis Publishers New York, 1-165.

Jenkins D., Richard M.G., Daigger G.T. (1993). Manual on the causes and control of activated sludge bulking and foaming. 2nd edn. Lewis Publishers New York, 1- 193.

Jenkins D., Richard M.G., Daigger G.T. (2003). Manual on the causes and control of activated sludge bulking and foaming. 3th edn. Lewis Publishers New York, 1- 190.

Jetten M.S., Sliekers O., Kuypers M., Dalsgaard T., van Niftrik L., Cirpus I., van de Pas-Schoonen K., Lavik G., Thamdrup B., Le Paslier D., Op den Camp H.J., Hulth S., Nielsen L.P., Abma W., Third K., Engstrom P., Kuenen J.G., Jorgensen B.B., Canfield D.E., Sinninghe Damste J.S., Revsbech N.P., Fuerst J., Weissenbach J., Wagner M., Schmidt I., Schmid M., Strous M.. (2003). Anaerobic ammonium oxidation by marine and freshwater planctomycete-like bacteria *App Microbiol Biot*, **63**, 107 – 114.

Juretschko S., Timmerman G., Schmid M., Schleifer K.H., Pommerening-Roser A., Koops H.P., Wagner M. (1998). Combined molecular and conventional analyses of nitrifying bacterium diversity in activated sludge: *Nitrosococcus mobilis* and *Nitrospira*-like bacteria as dominant populations. *Appl Environ Microbiol*, **64**, 3042-3051.

Krul J.M. (1977). Experiments with *Haliscomenobacter hydrossis* in continuous culture without and with *Zooglea ramigera*. *Water Res*, **11**, 197-204.

Larkin J. M., Shinabarger D. L. (1983). Characterization of *Thiothrix nivea*. *Int J Syst Bacteriol*, **33**, 841-846.

Levantesi C., Beimfohr C., Geurkink B., Rossetti S., Thelen K., Krooneman J., Snaidr J., van der Waarde J., Tandoi V. (2004). Filamentous Alphaproteobacteria associated with bulking in industrial wastewater treatment plants. *Syst Appl Microbiol*, **27**, 716-27.

Liu J. R., Seviour R. J. (2001). Design and application of oligonucleotide probes for fluorescent in situ identification of the filamentous bacterial morphotype in activated sludge. *Environ Microbiol*, **3**, 551-560.

Loy A., Horn M., Wagner M. (2003). ProbeBase: an online resource for rRNA-targeted oligonucleotide probes. *Nucl Acids Res*, **31**, 514-516.

Madoni P. (1994). *La microfauna nell'analisi di qualità biologica dei fanghi attivi. Indice biotico dei fanghi*. Azienda Gas Acqua Consorziale di Reggio Emilia - Università degli Studi di Parma, 1-48.

Mobarry B.K., Wagner M., Urbain V., Rittmann B.E., Stahl D.A. (1996). Phylogenetic probes for analyzing abundance and spatial organization of nitrifying bacteria. *App Environm Microb*, **62**, 2156-2162.

Mino T., Arun V., Tsuzuki Y., Matsuo T. (1987). Effect of phosphorus accumulation on acetate metabolism in the biological phosphorus removal process. In *Biological Phosphate Removal from Wastewaters*. Ramadori R. (ed). Oxford: Pergamon Press, 27-38.

Neef A., Zaglauer A., Meier H., Amann R., Lemmer H., Schleifer K.H. (1996). Population analysis in a denitrifiying sand filter: conventional and in situ identification of *Paracoccus spp.* in methanol-fed biofilms. *App Environm Microb*, **62**, 4329-4339.

Nielsen A.T., Liu W.T., Filipe C., Grady L., Molin S., Stahl D.A. (1999). Identification of a novel group of bacteria in sludge from deteriorated biological phosphorus removal reactor. *Appl Environ Microbiol*, **65**, 1251-1258.

Nielsen P.H., Roslev P., Dueholm T., Nielsen J.L. (2002). *Microthrix parvicella* a specialised lipid consumer in anaerobic-aerobic activated sludge plants. *Water Sci & Technol,* **46**(1-2), 73-80.

Odintsova E.V., Wood A.P., Kelly D.P. (1993). Chemolithoautotrophic growth of *Thiothrix ramosa Arch Microbiol*, **160**, 152-157.

Pereira H., Lemos P.C., Carrondo M.J.T., Crespo J.P.S.G., Reis M.A.M., Santos H. (1996). Model for carbon metabolism in biological phosphorus removal processes based on in vivo ^{13}C-NMR labelling experiments. *Wat Res*, **30**, 2128-2138.

Pike E.B. (1972). Aerobic bacteria in *Ecology Aspects of Used Water Treatment*. Curds C.R., Hawkes H.A. Eds., Academic Press, N.Y., 1-63.

Pitt P.A., Jenkins D. (1990). Causes and control of *Nocardia* in activated sludge. *Res J Water Pollu Control Fedn*, **62**, 143-150.

Polz M. F., Odintsova E.V., Cavanaugh C.M. (1996). Phylogenetic relationships of the filamentous sulfur bacterium *Thiothrix ramosa* based on 16S rRNA sequence analysis. *Int J Syst Bacteriol*, **46**, 94-97.

Richard M.G., Shimizu G.P., Jenkins D. (1985). Growth kinetics of *Sphaerotilus* species and their significance in activated sludge bulking. *JWPCF*, **57**, 68-81.

Rossetti S., Christensson C., Blackall L.L., Tandoi V. (1997b). Phenotypic and phylogenetic description of an Italian isolate of *M. parvicella. J Appl Bacteriol*, 82, 405-410.

Rossetti S., Tomei M.C., Levantesi C., Ramadori R., Tandoi V. (2002). *M. parvicella*: a new approach for kinetic and physiological characterization, *Water Sci & Technol*, **46**(1-2), 65-72.

Rossetti S., Blackall L.L., Levantesi C., Uccelletti D., Tandoi V. (2003). Phylogenetic and physiological characterization of a heterotrophic chemolithoautotrophic *Thiothrix* strain isolated from activated sludge. *Int J Syst Evol Microbiol*, **53**, 1271-1276.

Rossetti S., Tomei M.C., Nielsen P.H., Tandoi V. (2005). "*M. parvicella*" a filamentous bacterium causing bulking and foaming in activated sludge systems: a review of current knowledge. *FEMS Microbiol Rev*, **29**, 49-64.

Satoh H., Mino T., Matsuo T. (1994). Deterioration of enhanced biological phosphorus removal by the domination of microorganisms without polyphosphate accumulation. *Water Sci & Technol*, **30**(6), 203-211.

Schleifer K. H., Amann R., Ludwig W., Rothemund C., Springer N.S., Dorn (1992). Nucleic acid probes for the identification and in situ detection of pseudomonads In Galli E., Silver S., Witholt B. Eds. *Pseudomonas: molecular biology and biotechnology* American Society for Microbiology. Washington D.C. 127-134.

Schramm A., de Beer D., van den Heuvel J.C., Ottengraf S., Amann R. (1999). Microscale distribution of populations and activities of *Nitrosospira* and *Nitrospira* spp. along a macroscale gradient in a nitrifying bioreactor: quantification by *in situ* hybridization and the use of microsensors. *Appl Environ Microbiol*, **65**, 3690-3696.

Seviour E.M., Seviour R.J., Lindrea K.C. (1999). Description of the filamentous bacteria causing bulking and foaming in activated sludge plants. In: The Microbiology of activated sludge. Seviour R.J., Blackall L.L Eds., 1-422, Kluwer Academic, London.

Seviour R.J. Maszenan AM Soddell JA Tandoi V. Patel BK Kong Y Schumann P. (2000). Microbiology of the 'G-bacteria' in activated sludge. *Environ Microbiol*, **2**, 581-93.

Seviour RJ Liu J.R. Seviour E.M. McKenzie C.A. Blackall L.L. Saint C.P. (2002). The "Nostocoida limicola" story: resolving the phylogeny of this morphotype responsible for bulking in activated sludge. *Water Sci & Technol*, **46**(1-2), 105–110.

Seviour R.J., Mino T., Onuki M. (2003). The microbiology of biological phosphorus removal in activated sludge systems. *FEMS Microbiol Rev*, **27**, 99-127.

Slijkhuis H., Deinema M.H. (1982). The physiology of *M. parvicella* a filamentous organism isolated from activated sludge In *Bulking in Activated Sludge: Preventative and Remedial Methods*. Chambers B., Tomlinson E.J. Eds.. Chichester, Ellis Horwood Ltd. 75-83.

Snaidr J., Beimfohr C., Levantesi C., Rossetti S., van der Waarde J., Geurkink B., Eikelboom D., Lemaitre M., Tandoi V. (2002). Phylogenetic analysis and in situ identification of Nostocoida limicola-like filamentous bacteria in activated sludge from industrial wastewater treatment plants. *Water Sci & Technol*, **46**(1-2), 99–104

Schramm A., de Beer D., Wagner M., Amann R. (1998). Identification and activities in situ of *Nitrosospira* and *Nitrospira* spp. as dominant populations in a nitrifying fluidized bed reactor. *Appl Environm Microbiol*, **64**, 3480-3485.

Tandoi V., Caravaglio N., Di Dio Balsamo D., Majone M., Tomei M.C. (1994). Isolation and physiological characterization of *Thiothrix* sp.. *Water Sci & Technol*, **29**(7), 261-269.

Tandoi V., Majone M., May J., Ramadori R. (1998b). The behaviour of polyphosphate accumulating *Acinetobacter* isolates in an anaerobic-aerobic chemostat. *Water Res*, **32**, 2903-2912.

Tandoi V., Rossetti S., Blackall L.L., Majone M. (1998a). Some physiological properties of an Italian isolate of *M. parvicella*. *Water Sci & Technol*, **37**(4-5), 1-8.

van Niekerk A.M., Jenkins D., Richard M.G. (1987). The competitive growth of *Zooglea ramigera* and Type 021N in activated sludge and in pure culture - A model for low F:M bulking, *JWPCF*, **59**, 262-273.

van Veen W.L. (1973). Bacteriology of activated sludge in particular the filamentous bacteria. *Antonie van Leeuwenhoek*, **39**, 189-205.

Wagner M., Amann R.I., Kampfer B., Assmus B., Hartmann A., Hutzler P., Springer N., Schleifer K.H. (1994a). Identification and in situ detection of Gram-negative filamentous bacteria in activated sludge. *Syst Appl Microbiol*, **17**, 405-417.

Wagner M., Erhart R., Manz W., Amann R., Lemmer H., Wedi D., Schleifer K.H. (1994b). Development of an rRNA-targeted oligonucleotide probe specific for the genus *Acinetobacter* and its application for in situ monitoring in activated sludge. *Appl Environ Microb*, **60**, 792-800.

Wagner M., Rath G., Amann R., Koops H.P., Schleifer K.H. (1995). In situ identification of ammonia-oxidizing bacteria. *Syst Appl Microbiol*, **18**, 251-264.

Williams T. M., Unz R.F. (1985). Isolation and characterization of filamentous bacteria present in bulking activated sludge. *App Microbio Biot*, **22**, 273-282.

Woese C.R. (1987). Bacterial evolution. *Microb Rev*, **51**, 221-271.

3

AS Separation problems

Jiri Wanner

3.1 INTRODUCTION

The activated sludge process is the most widespread method for the biological treatment of both municipal and industrial wastewaters. It is used in many process modifications and at scales that range from small package units for individual dwellings in rural areas up to large installations for big cities and industrial complexes. In all cases, the biologically treated wastewater must be separated from the activated sludge biomass to produce a high-quality final effluent. Although new separation processes like membrane filtration have been tested and used to a limited extent in pilot- and full-scale units, the gravity sedimentation is still the only economically feasible solids separation method for most municipalities and industrial enterprises. Because of this the importance of having good activated sludge settling properties cannot be overemphasized.

Activated sludge bulking caused by filamentous microorganisms is the most common operating problem associated with poor activated sludge separation. Discussions of activated sludge bulking are found frequently in papers, books and practical manuals on wastewater treatment. Because of this many practitioners tend to use the term " activated sludge bulking" as a synonym for all activated sludge separation problems. This misinterpretation can lead to inappropriate control measures.

3.2 "WELL-SETTLING" ACTIVATED SLUDGE

3.2.1 Requirements

For successful wastewater treatment using activated sludge the biomass must be able to (i) flocculate and (ii) settle and thicken by gravity sedimentation because the treated wastewater is separated from the activated sludge in secondary clarifiers (secondary settling tanks) where the main driving force for separation is gravitation. A large and compact activated sludge floc (see Figure 3.1) is a prerequisite for efficient activated sludge separation in secondary clarifiers. Poor separation and thickening of activated sludge in secondary clarifiers can lead to the following operating problems:

- Activated sludge biomass escapes from the secondary clarifier and deteriorates final effluent quality not only in the terms of suspended solids, but also in BOD_5 (respiration of biomass), COD (organic nature of biomass) and in TKN and TP (the biomass content of N and P)
- A dilute return sludge stream is produced that may not allow maintenance of the desired aeration basin biomass concentration and proper control of sludge age
- A dilute waste sludge stream hydraulically overloads the sludge handling facilities

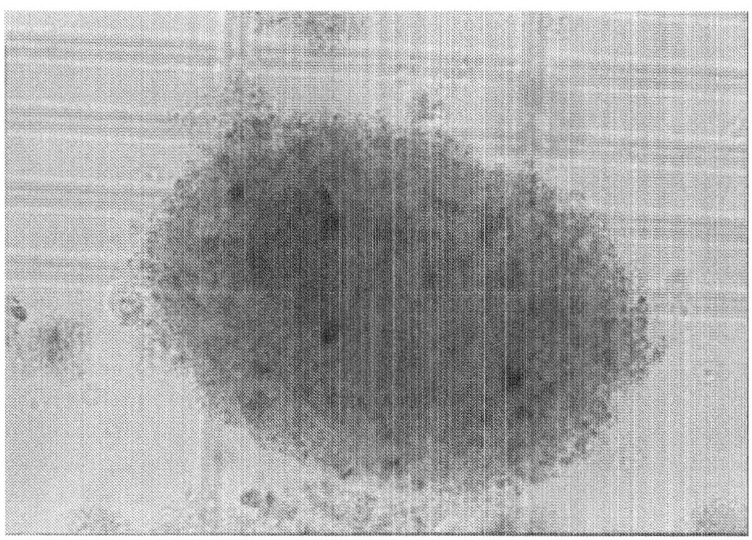

Figure 3.1. Large and compact activated sludge floc (250X, wet mount , phase contrast)

Well settling activated sludge should have the following properties (Wanner, 1994):

- It settles fast, with zone settling velocities of >3 m/h
- It does not occupy an excessive volume after settling and thickening in a secondary clarifier
- After sedimentation it leaves a clear supernatant (X = 15 mg/L or less)
- It does not rise after sedimentation for at least 2-3 h

3.2.1.1 Measurement of settling and thickening properties

Conditions during settling in secondary settling tanks are best simulated by zone settling velocity measurements. The height of the sludge interface is measured over time in as tall and wide a cylinder as possible. The plot of sludge layer height vs. time is called a settling curve. This curve usually has three distinct settling phases (see Figure 3.2):

I - reflocculation
II - zone settling
III, IV - transition and compaction

The zone settling velocity is calculated from the slope of the curve in the zone settling phase (II).

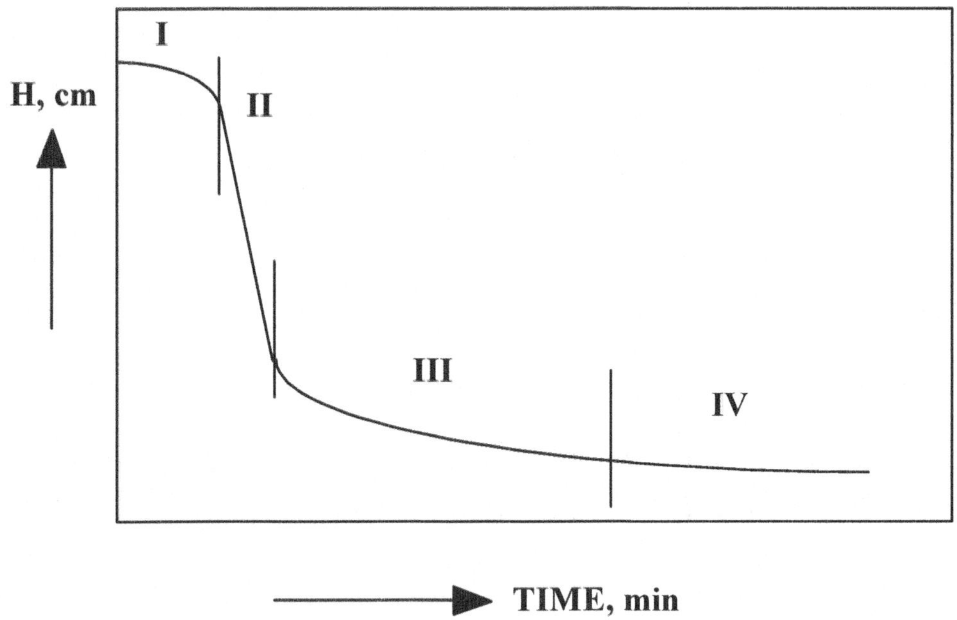

Figure 3.2. An example of activated sludge settling curve

In wastewater treatment practice the commonest way to measure activated sludge settleability is by the sludge volume index SVI test. The standard sludge volume index is defined as follows:

$$SVI = \frac{V_{30}}{X} \qquad \text{(eq. 3.1)}$$

where

- SVI = sludge volume index [mL/g]
- V_{30} = volume of settled sludge after 30 min sedimentation in 1L cylinder
- X = concentration of activated sludge prior to settling (mixed liquor suspended solids) [g/L]

SVI values are affected by many factors. Especially important are the effects of sludge concentration or the volume of settled sludge after 30 min and wall effects in the cylinder. In some countries settling test conditions are standardized. Common standardization methods include:

- Stirred sludge volume index SSVI; the measuring cylinder is equipped with slowly rotating stirring device
- SVI at a standard suspended solids concentration; the test is performed with a defined sludge concentration (for instance 3.5 g L, $SVI_{3.5}$)
- Diluted sludge volume index DSVI; the volume of settled sludge after 30 min sedimentation should not exceed 200 mL/L. If this occurs the test is repeated with a diluted sludge

Activated sludge can be classified on the basis of its zone settling velocity (ZSV) and sludge volume index as shown in Table 3.1.

Table 3.1. Activated sludge settleability classification

Type of sludge	SVI, mL/g	ZSV, m/h
Well settling	< 100	> 3
"Light"	100-200	2-3
Bulking	> 200	< 1.2

When the sludge separation problems are caused by filamentous microorganisms, it is useful to quantify the amount of these microorganisms in activated sludge by a method such as that of Sezgin et al. (1978), where total length of filaments protruding from flocs or freely floating in the bulk liquid(TEFL) are measured. TEFL values of 10^7 µm/mL (10^4 m/g) are considered to represent the transition point between a non-bulking and a bulking activated sludge.

Because the measurement of TEFL is laborious and time consuming so for the routine examination of activated sludge, a rapid, simple method for subjective scoring of filament abundance was developed by Jenkins et al. (1986). The abundance of filaments is classified according to the scale given in Table 3.2.

Table 3.2. Subjective scoring of filament abundance according to Jenkins et al. (1986)

Numerical value	Abundance	Description of microscopic picture
0	none	
1	few	Filaments present, but only observed in an occasional floc
2	some	Filaments commonly observed, but not present in all flocs
3	common	Filaments observed in all flocs, but at low density (e.g, 1-5 filaments per floc)
4	very common	Filaments observed in all flocs at medium density (e.g., 5-10 filaments per floc)
5	abundant	Filaments observed in all flocs at high density (e.g., > 20 filaments per floc)
6	excessive	Filaments present in all flocs - appears more filaments than flocs and or filaments growing in high abundance in bulk solution

For the foam-forming nocardioform actinomycetes, which are often irregularly-shaped and frequently branched, Pitt and Jenkins (1990) recommend a counting method based on determining the number of intersections of these filaments with a hairline in the eyepiece (Chapt 2.12).

3.2.2 Microscopic features of well settling activated sludge

A simple microscopic examination can be very useful for the assessment of sludge separation problems. Good settling sludge is formed of roughly spherical, compact, firm flocs (Figure 3.1) With an average size of > 100 µm. Phase contrast microscopic observation of such flocs at 1000 magnifications will often reveal a "filamentous backbone". This "filamentous backbone" can be better seen in Gram or Neisser stained samples. This suggests that large compact flocs are formed by bacterial growth on the filaments observed inside the floc. When the filaments protrude from the flocs into the bulk liquid they can cause *bulking sludge* (see below). This is referred to as poor floc *macrostructure.* In some activated sludge samples, the compact core of the flocs created by bacterial biomass is absent and the flocs exhibit a loose, open structure. When the compact core of activated sludge flocs is too small (usually 20-80 µm) the cause for poor settling is said to be due to poor floc *microstructure.* Such microstructure problems can be due either to insufficient production or overproduction of the biopolymer material (glycolax) that binds the cells together in of the activated sludge flocs.

3.3 ACTIVATED SLUDGE SEPARATION PROBLEMS

Six major activated sludge operating can be problems related to microbial biomass quality (Jenkins et al., 1986; Wanner, 1994; Wanner, 2002):

3.3.1 Poor floc microstructure

The following problems depends on a floc microstructure failure.

3.3.1.1 Dispersed growth

The activated sludge microorganisms are dispersed freely in the bulk liquid as individual cells (Figure 3.3) or small clumps with diameters of up 10-20 µm. The sedimentation rate of these individual cells or bacterial clumps is too low for them to be removed efficiently by gravity sedimentation. Also no zone settling occurs in secondary settling tanks. This has two impacts on the activated sludge process:

- The separation efficiency of the secondary settling tank is very poor and the final effluent is turbid
- Because of the poor solids separation efficiency, a significant amount of biomass escapes from the system so that only low sludge ages (Θ_X) can be maintained. A system with dispersed growth is closer to a chemostat than a continuous cultivation system with biomass recycle.

The poor bioflocculation that is the root cause of dispersed growth is caused by low production of the extracellular biopolymers that create the matrix in firm activated sludge flocs. A common cause of dispersed growth is a very high organic loading (high "food to microorganism" ratio, F/M) since bacteria do not need to produce a glycocalyx under these conditions.Dispersed growth can be a common problem during the start-up of activated sludge systems. Another cause is the presence of toxicity in the wastewater.

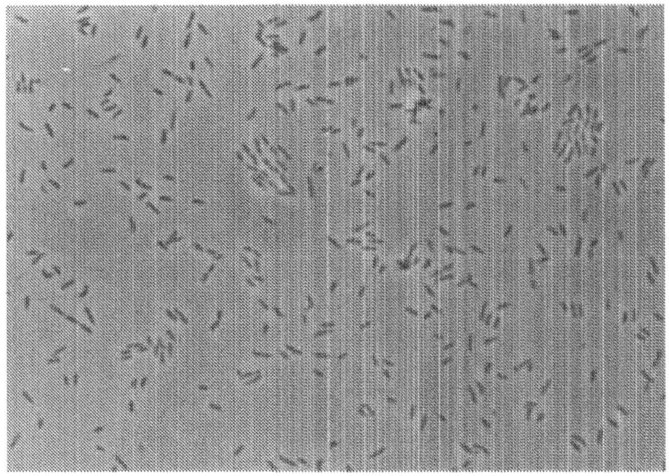

Figure 3.3. Dispersed freely growing bacterial cells (250X, wet mount, phase contrast).

3.3.1.2 Unsettleable microflocs

The appearance that this separation problem gives is quite similar to that of dispersed growth. The final effluent from secondary clarifier is not very clear and it contains many small particles of escaping biomass (Figured 3.4). However microscopic examination will reveal that the nature of the problem is different. The unsettleable particles are larger than in dispersed growth (about 50-100 μm) and are roughly spherical and compact. These microflocs result from the disintegration of firm and sound flocs.

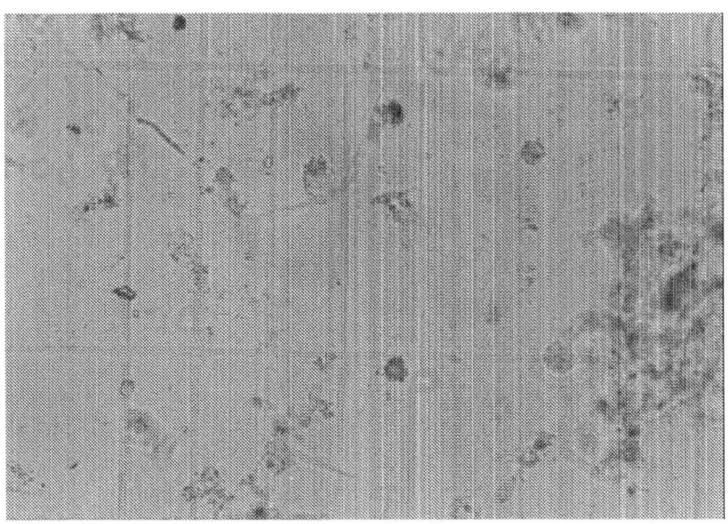

Figure 3.4. Unsettleable microflocs of activated sludge (250X, wet mount, phase contrast).

During the settling test the activated sludge rapidly separates into two fractions: large flocs and small flocs. The large flocs settle rapidly and when the SVI is calculated on the basis of the volume of these larger flocs, its value is quite low (around 50 mL/g). The supernatant in the cylinder is turbid because of the suspended small flocs and a substantial fraction of the total biomass remains in these unsettleable particles.

Reasons for floc disintegration are:

- Insufficient production of glycocalyx or its consumption by bacteria inside the flocs at low organic loadings (typical of high sludge age, extended aeration systems)
- Total absence of filamentous microorganisms which form the "backbones" of larger flocs (Jenkins et al., 1986)
- Disintegration of flocs by shearing effects, for instance by high speed mechanical aerators

3.3.1.3 Viscous bulking

The rather broad term viscous bulking describes symptoms rather than causes. Viscous bulking activated sludge contains an excessive amount of extracellular biopolymers which impart a slimy, jelly-like consistency to it. Because the biopolymers are hydrophilic the activated sludge becomes highly water retentive. Such a "hydrous" activated sludge exhibits low settling and compaction velocities. Also since the biopolymers are surface active, a viscous activated sludge may foam when aerated intensively (Wanner, 2002).

Biopolymer production is characteristic of most floc forming microorganisms. Under normal conditions (no toxic compounds, nutrient balanced growth) the amount of biopolymer generated is just enough for the formation of firm flocs. Zoogloeal bacteria always produce large amounts of biopolymers because their individual cells are surrounded by a significant slime layer. Because of this some authors use the term "zoogloeal bulking" to describing the settling problems caused by the presence of an excessive amount zoogloeal colonies in activated sludge (Figures 3.5 and 3.6).

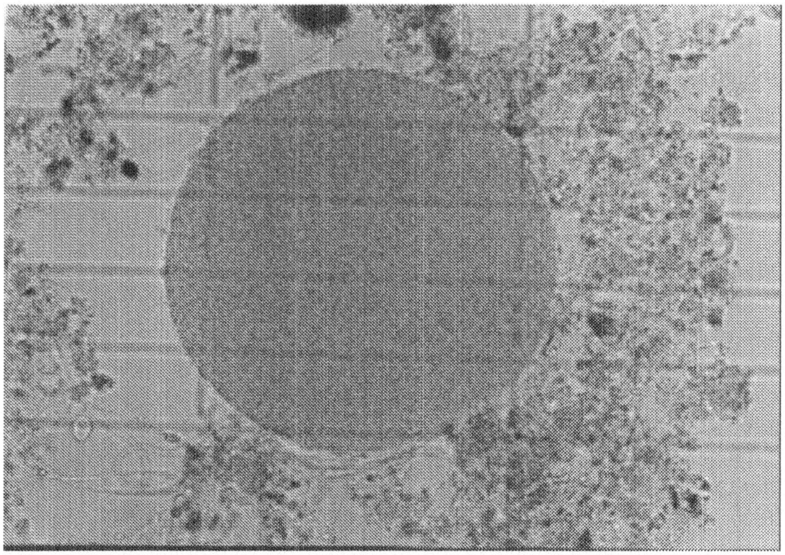

Figure 3.5. Large amorphous colony of zoogloeal bacteria (250X, wet mount, phase contrast).

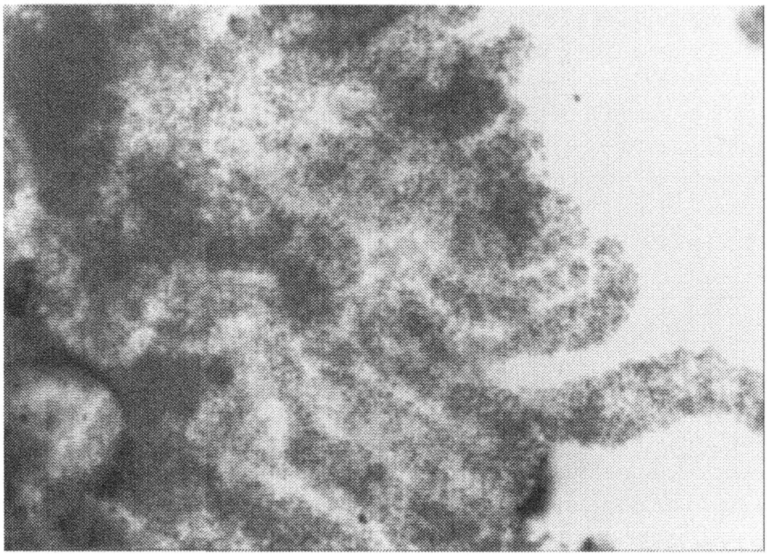

Figure 3.6. Large fingered colony of zoogloeal bacteria (1000X, Gram stain, direct illumination).

3.3.2 Poor floc macrostructure

The following problems depend on a floc macrostructure failure.

3.3.2.1 Filamentous bulking

Filamentous bulking is a common case of poor activated sludge compaction (phase III of the settling curve – see above) which results in (Wanner, 1994):

- Low return and wasted activated sludge solids concentrations
- Difficulties in maintaining the required activated sludge concentration in aeration basins
- Poor sludge dewaterability
- Hydraulic overloading of sludge handling facilities

Filamentous microorganisms interfere with the sedimentation and compaction of activated sludge flocs in two ways (Jenkins et al., 1986):

1. Some filamentous microorganisms grow inside flocs giving them a diffuse open structure. These open flocs have space to retain water in them, so that although aggregation of individual flocs is not mechanically hindered by filaments protruding from the flocs, excessive water remains "captured" in the settled sludge.
2. Filaments can detrimentally affect the sedimentation and compaction of activated sludge flocs, that is much more common. Most of filamentous microorganisms observed in activated sludges protrude from rather compact and firm flocs into the bulk liquid. These filaments, which in low numbers can form the backbones of firm flocs, in large numbers are able to mechanically prevent the compaction of individual flocs. This type of interference problem is called “bridging” (Figure 3.7).

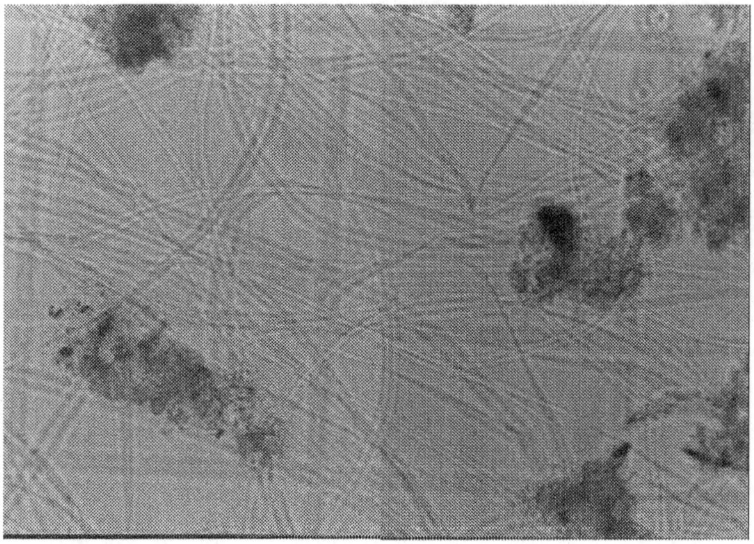

Figure 3.7. Filamentous microorganism growing in excess between activated sludge flocs (250X, wet mount, phase contrast).

3.3.2.2 Foaming caused by filamentous microorganisms

Biological foaming by "foam-forming" filamentous microorganisms is a combination of physico-chemical and biochemical interactions leading to the stabilization of a three-phase system of air - water - microbial cells. Stabilization of these biological foams is caused by the following features of foam-forming filaments:

- Production of extracellular materials such as lipids, lipopeptides, proteins and carbohydrates which have surface active properties (biosurfactants)
- The cell walls of foam-forming microorganisms are strongly hydrophobic

The formation of stable foams in the aeration basins of activated sludge plants can create several operating problems:

1. aesthetic problems, such as slippery pathways covered by escaping foam
2. floating biomass in the secondary clarifier that can deteriorate final effluent quality
3. accumulation of significant amount of biomass into the foam that is not active in the treatment processes; loss of ability to control activated sludge age

Biological foam, especially on the surface of a secondary clarifier, may look like floating activated sludge. To distinguish between biological foam and floating sludge, microscopic examination can be very useful. The abundance of filamentous microorganisms in biological foams is usually much higher than in the corresponding mixed liquor. On the contrary, the microscopic appearance of floating sludge is almost the same as the mixed liquor (unless bubbles of nitrogen gas are present). Especially when caused by *M. parvicella,* the surface foam on the aeration tanks, and secondary clarifiers can look like a mono-culture of the filament (see Figure 3.8).

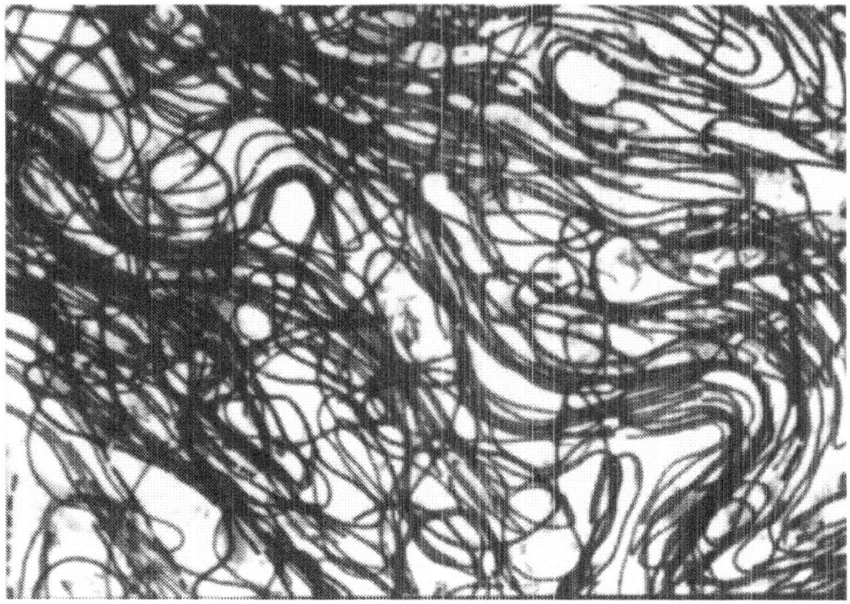

Figure 3.8. Microscopic picture of biological foam caused by Microthrix parvicella (1000X, Gram stain, direct illumination).

There are no unified and generally accepted methods for the quantification of foaming problems. Probably the most exact method for predicting the amount of foam formed on aeration basins and the extent of the resulting problems is the Scum Index (SI) suggested by Pretorius and Laubscher (1987). SI [%] is calculated as follows:

$$SI = \frac{\text{mass of biomass in foam}}{\text{total mass of biomass}} \times 100 \qquad \text{(eq. 3.2)}$$

The portion of the biomass in the foam is estimated by floatation using a standard aeration rate of 10 m^3/m^3 h in in a batch floatation cell (80 mm diameter; 500 mm high). The floatation is repeated several times until all foam-forming microorganisms have been transferred into the scum. SI can be used for predicting the operating problems that might occur using the information presented in Table 3.3

Table 3.3 Scum index and expected operational problems according to Pretorius and Laubscher (1987)

SI %	Extent of Problem
0-0.5	Negligible
0.5-6	Small
6-10	Medium
10-15	Serious
> 15	Catastrophic

Plant operators can monitor the tendency of activated sludge to foam using the method suggested by Kocianova et al. (1992). The percentage of aeration basin and secondary clarifier surfaces covered by foam is regularly recorded. The observations should be made at the same time of the day when loading conditions and operating procedures are similar. The authors found a good correlation between % coverage and other parameters reflecting the presence of foam-formers such as foam accumulation rate and the hydrophobicity of the activated sludge.

Recently, the measurement of hydrophobicity has become more common because this parameter is directly connected with the phenomenon of biological foaming. The hydrophobicity of mixed liquor and foam can be measured by using the method of Rosenberg et al. (1980) as modified by Kocianova et al. (1992) – the so-called MATH test (Microbial Adherence To Hydrocarbons). The assay is based on adhesion of hydrophobic microorganisms to a hydrophobic surface of a solvent like hexadecane. The results of the MATH assay as modified by Kocianova et al. (1992) are affected by the electrical surface charge properties of the water/hexadecane system because the droplets of hexadecane formed during the "extraction" of hydrophobic microorganisms from the aqueous phase are highly negative (Medrzycka, 1991), presumably due to the adsorption of anions.

Foaming intensity depends strongly on the concentration of surface active compounds. Nowadays products are containing non-ionic detergents are used both in households and in industry. The concentrations of non-ionic detergents are not usually measured by wastewater laboratories because of the complexity of the analytical procedures. Because of this the concentration of non-ionic detergents in the influent wastewater should be checked when some unexpected foaming occurs.

Extracellular polymers (ECPs) produced under certain conditions by activated sludge microorganisms can be also responsible for the formation of biological foam, especially in aeration basins. The presence of ECPs is not measured commonly in activated sludge. If frequent foaming periods occur it is recommended that the measurement of ECPs should be prerformed regularly. If these show a tendency to increase in concentration some operational counter measures should be taken, e.g., balancing the influent BOD with nitrogen and/or phosphorus or checking to determine whether toxic compounds are present in the raw wastewater. The analytical procedures for the determination of carbohydrates, proteins and DNA (the main constituents of ECPs) can be found in specialized literature.

3.3.3 Other reasons

3.3.3.1 Rising sludge

When this is the problem, the water surface in the secondary clarifier is covered by patches (or in severe cases complete coverage) of floating activated sludge. If the phenomenon is observed in a glass cylinder, two phases can be distinguished:

- The activated sludge settles rapidly and a rather compact bottom layer of settled sludge and a clear supernatant are formed
- After a certain period of time part or a all of the settled and thickened sludge starts to float and move up to the water surface. (If this occurs within 30 min it can interfere with the SVI measurement)

The floating material is full of gas bubbles derived from denitrification which takes place in the settled and thickened layer of activated sludge. Because the settled sludge has a high biomass concentration, the dissolved oxygen is quickly depleted and if the activated sludge is from a nitrifying plant, anoxic conditions are established. The bubbles of nitrogen liberated during denitrification act as sludge "carriers". Thus, the primary reason for this kind of separation problem does not lie in the activated sludge itself but in:

- Insufficient denitrification in the activated sludge system

- Poor performance of settled sludge removing equipment in secondary clarifiers that allows the sludge to remain too long in the clarifier.

3.4 SUMMARY

The primary conditions for successful activated sludge system performance are good separation and thickening properties of activated sludge. The quality of the final effluent from an activated sludge system is determined by the efficiency of activated sludge separation in secondary clarifiers. Six types of activated sludge separation problem can cause the deterioration of secondary clarifier final effluent. Some practitioners used to call all these separation problems "activated sludge bulking". This is incorrect and does not allow to determine the casues of a given separation problem or to choose a proper control method. This section has described the characteristic features of individual separation problems. The importance of microscopic examination of activated sludge for the identification of separation problems is stressed and has been illustrated by microphotographs. Quantification of separation problems together with careful microscopic examination of activated sludge should always be the first step in solving and rectifying activated sludge separation properties. For foaming problems the conventional methods of quantification of separation problems like SVI and ZSV measurement should be accompanied by the measurement of biomass hydrophobicity and by the estimation of the of non-ionic detergent and extracellular polymer concentrations.

Acknoweledgment: Microphotographs were taken by Olga Krhutkova.

3.5 REFERENCES

Jenkins D., Richard M.G., Daigger G.T. (1993). Manual on the Causes and Control of Activated Sludge Bulking and Foaming. 2nd Edition CRC Press, Lewis Publishers, Boca Raton (FL,) USA.

Kocianova E., Foot R.J., Forster C. F. (1992). Physicochemical aspects of activated sludge in relation to stable foam formation. *J Inst Water Environ Manage*, **6**, 342-350.

Medrzycka K. B. (1991). The effect of particle concentration on the zeta potential in extremely dilute solutions. *Coll Pol Sci*, **269**, 85-90.

Pitt P., Jenkins D. (1990). Causes and control of *Nocardia* in activated sludge. *Res J Water Pollut Control Fed*, **62**, 143-150.

Pretorius W.A., Läubscher C.J.P. (1987). Control of biological scum in activated sludge plant by means of selective flotation. *Water Sci & Technol,* **19**(5-6). 1003-1011.

Rosenberg M. Gutnick D. Rosenberg E. (1980). Adherence of bacteria to hydrocarbons: A simple method for measuring cell-surface hydrophobicity. *FEMS Microbiol Lett*, **9**, 29-33.

Sezgin M., Jenkins D., Parker D.S. (1978). A unified theory of filamentous activated sludge bulking. *J Water Pollut Control Fed*, **50**, 362-381.

Wanner J. (1994). *Activated Sludge Bulking and Foaming Control.* Technomic Publishing Co. Inc. Lancaster (PA), USA.

Wanner J. (2002). Control of filamentous bulking in activated sludge. In: *Encyclopedia of Environmental Microbiology*, Bitton G. Ed., 1306-1315. John Wiley & Sons Inc. New York USA.

4

Aeration tank and secondary clarifier as one system

J. Wanner

4.1 INTRODUCTION

Traditionally the activated sludge process is treated as a two stage process. The first stage (aeration basin) is described in terms of biological processes while the description of the second stage concentrates on processes of biomass separation and thickening. The modern approach to the activated sludge process looks at both the biological and the separation stage as one system. The aim of this chapter is to explain the interactions between the two parts of the activated sludge system. For information on the principles of design and construction of secondary clarifiers. the reader should consult the basic literature on secondary clarifiers from recent years (Albertson, 1992; ATV – DVWK Standard, 2000; Ekama et al., 1997). The activated sludge system, comprising a biological reactor and a separation/thickening stage, can be described schematically by Figure 4.1.

 Activated Sludge Separation Problems: Theory, control measures and practical experience edited by V. Tandoi, D. Jenkins And J. Wanner. ISBN: 1900222841. Published by IWA Publishing, London, UK.

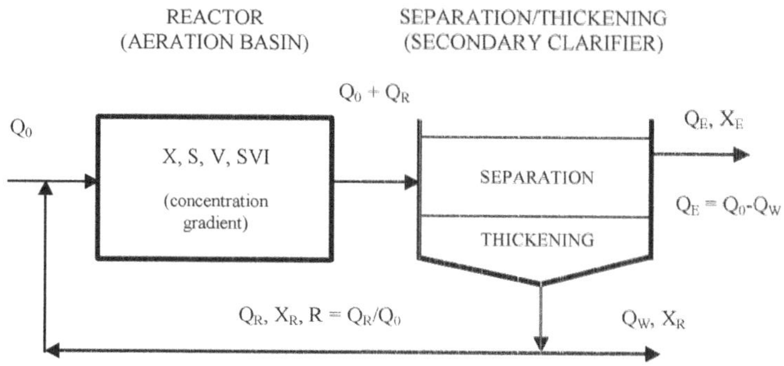

Figure 4.1 Simplified flow-scheme of an activated sludge system

4.2 AERATION TANK AND SECONDARY CLARIFIER INTERACTIONS

4.2.1 Activated sludge process

The major technological parameter affecting the performance of the activated process in the reactor (aeration basin) is the biomass retention time, or sludge age, defined as follows:

$$\Theta_X = \frac{V \cdot X + M_{FC}}{Q_W \cdot X_R + Q_E \cdot X_E} \qquad \text{(eq. 4.1)}$$

Where:

Θ_X	biomass retention time, sludge age (d)
M_{FC}	amount of activated sludge biomass stored in secondary clarifier (kg)
V	reactor volume (m^3)
X	biomass concentration in the reactor (mg/L)
Q_W	waste sludge flow rate (m^3/d)
X_R	concentration of thickened sludge (mg/L)
Q_E	effluent flow rate (m^3/d)
X_E	effluent biomass concentration (mg/L)

Equation (4.1) shows that the biomass retention time in the entire activated sludge system is directly affected by:

- the storage capacity of the secondary clarifier M_{FC} (dependent on the construction of the clarifier and its operating mode)
- the concentration of thickened sludge X_R which is a function of both activated sludge quality and the design and construction of the secondary clarifier
- the concentration of biomass in the effluent X_E which is partially influenced by the activated sludge separation characteristics (see Chapter 3) and by the construction and loading of the secondary clarifier.

The performance of the secondary clarifier can also affect biomass retention time indirectly i.e., through the mixed liquor concentration (X) that it is possible to maintain

$$X = \frac{R}{1+R} \cdot X_R \qquad \text{(eq. 4.2)}$$

where R is the return sludge ratio.

The concentration of thickened sludge, X_R, depends both on the activated sludge settling properties and on the performance of the secondary clarifier. In the past, the German design procedure A131 used Kalbskopf 's simple relationship of:

$$X_R = \frac{1200}{SVI} \qquad \text{(eq. 4.3)}$$

(X_R in g/L and SVI in mL/g)

Today, more sophisticated relationships for the prediction of recycle flow concentration X_R can be found the literature (Ekama et al., 1997).

In general, the poor thickening in the secondary clarifier leads to problems with:

- control of biomass retention time (see above)
- hydraulic overloading of sludge handling facilities (because of low X_R the waste sludge flow rate Q_W has to be increased)

The loss of control over the sludge age Θ_X can cause various separation problems. Because many activated sludge systems are operated as BNR plants, problems with sludge age control may cause instabilities in nutrient removal processes, especially in nitrification.

4.2.2 Secondary clarifier

4.2.2.1 General performance, thickening function

The performance of a secondary clarifier is affected by three major parameters:

- q_A surface overflow rate $m^3/(m^2 \cdot h)$

$$q_A = \frac{Q_E}{A_{SC}} \qquad \text{(eq. 4.4)}$$

where A_{SC} is the surface area of secondary clarifier (m^2)

- Θ hydraulic retention time (h)
- N_A solids loading rate (applied flux) $kg/(m^2 \cdot h)$

$$N_A = \frac{X \cdot Q_0 \cdot (1+R)}{A_{SC}} \qquad \text{(eq. 4.5)}$$

where Q_0 is the influent flow rate (m^3/d)

Because these parameters are affected by the settleability of the activated sludge, the numerical values of clarifier design criteria, in various design manuals, are usually limited to a range of activated sludge quality. For instance, the Czech Design Recommendation CSN 75 6401 states that the design criteria are valid only for SVI less than 150 mL/g and for recycle ratios less than 1.5.

4.2.3 Separation function

Some activated sludge separation problems can also affect secondary clarifier effluent quality (X_E). Elevated effluent biomass concentration can result from:

- dispersed growth of activated sludge activated sludge or the formation of unsettleable microflocs
- rising sludge (denitrification) or floating biological foam when the effluent weir is not protected by baffles

Increases of biomass concentration X_E in the final effluent leads to deterioration of other effluent parameters.

4.2.4 BOD_5

The BOD_5 of the final effluent consists of two fractions: soluble and particulate. The particulate BOD_5 is caused by the respiration (mostly endogenous) of microorganisms which form the suspended solids in the final effluent.

$$(BOD_5)_{TOT} = (BOD_5)_{SOL} + k \cdot X \qquad \text{(eq. 4.6)}$$

The coefficient *k* in Equation (4.6) expresses the activity of biomass (x*)*. The amount of respiration depends on the organic fraction X_O of total suspended solids (which corresponds to the biomass fraction) and on the biomass retention time (Θ_X.) Based on a correlation of data from Czech activated sludge plants, the following empirical equation can be used for the determination of k (Chudoba *et al.*, 1991):

$$k = \frac{0.7 \cdot (X_O/X)}{1 + 0.044 \cdot \Theta_X} \qquad \text{(eq. 4.7)}$$

4.2.5 COD

The residual COD consists of two fractions – soluble and particulate. Soluble COD consists of biologically non-degradable compounds which:

- were either present in raw wastewater
- resulted from microbial activity of activated sludge microorganisms.

Most of the particulate COD results from the oxidation of biomass in the suspended solids (X_O) during the COD test. For a rough calculation of residual COD the following empirical relation can be used:

$$1 \text{ mg } X_O \approx 1.4 \text{ mg COD}$$

4.2.6 Nitrogen and phosphorus

The biomass fraction of activated sludge effluent suspended solids X contains 6–8 % organic nitrogen (measured as Kjeldahl nitrogen) and 1.5-2 % phosphorus for conventional systems, 4-6 % phosophorus for BNR plants incorporating enhanced biological phosphorus removal.

4.3 THE IMPACT OF AERATION BASIN EQUIPMENT AND OPERATION ON THE PERFORMANCE OF SECONDARY CLARIFIERS

4.3.1 Mechanical vs. diffused-air aeration

Both mechanical and diffused-air aeration systems are used for mixing and aerating activated sludge mixed liquor. The effect of various types of aerators on the quality of activated sludge had been a topic of discussion for many years. Konicek and Burdych (1988) showed that diffused air systems were superior to mechanical aeration by:

- microscopic pictures of activated sludge flocs in aeration basins
- a statistical evaluation of the relationship between total effluent suspended solids concentration X_E and type of aeration
- an evaluation of the size distribution of microparticles in the effluent from activated sludge systems aerated by both types of aeration devices.

In the 1990s, diffused-air aeration using fine bubble diffusers became common in new and upgraded activated sludge plants mostly because of economic reasons. These full-scale plants confirmed the results of Konicek and Burdych (1988). In most modern activated sludge systems the presence of activated sludge microflocs in secondary clarifier effluents does not result from the action of aeration system.

Figures 4.2, and 4.3 clearly demonstrate the difference between the turbulent environment created by mechanical aerators (4.2) and the mild conditions created by fine-bubbles diffused-air aerators (4.3).

Figure 4.2 Aeration by vertical-shaft mechanical aerator (aeration turbine)

Figure 4.3 Aeration by fine-bubble diffused-air system is friendly to flocs

4.3.2 Mixed liquor mixing

BNR activated sludge plants require non-aerated conditions (anaerobic and/or anoxic). To keep the activated sludge in suspension and in contact with the wastewater, these non-aerated reaction zones must be mechanically stirred. Nowadays it is common to use submerged mixers with both horizontal and vertical shafts. The primary consideration in the design of these mixers is the avoidance of excessive shear stresses during contact by the activated sludge flocs with the body of the propeller. Typically low-speed, large-diameter propellers are used. The shape of the propeller blades is such that maximum energy efficiency is provided without side-effects such as cavitation that would lead to floc disintegration.

4.3.3 Degasification and reflocculation in aeration tanks

Aeration tanks with fine-bubble diffused-air aerators can be rather deep. Depths around 4 m are most common but aeration tanks up to 12 m deep are becoming more and more common. While deep tanks can improve oxygen transfer efficiency they can also cause separation problems in the secondary clarifier because the mixed liquor entering the secondary clarifiers is supersaturated with air. On entering the shallower secondary clarifier, small air bubbles are released that attach to the activated sludge flocs and cause them to float. The result of this problem is similar to "rising sludge" (see Chapter 3.3.3.1).

The remedy for this problem is quite simple and consists of a degasification and reflocculation zone that can be incorporated directly into the aeration tank system. There are two approaches for the construction of degasification chambers. In the first case the activated sludge mixed liquor flows from the intensively aerated environment of the aeration basin in a very thin layer over a weir. The abrupt change of the hydrostatic pressure between the aeration basin and this thin layer causes a rapid release of air bubbles, that are then stripped-off by the coarse-bubble aeration in the degasification chamber. The gentle mixing by these large bubbles also contributes to the reflocculation of activated sludge flocs into aggregates larger than those in the aeration basin. The second approach, (Fig. 4.4) exploits the hydraulic mixing that results from the flow of mixed liquor. The degasification chamber is separated from the aeration basin by a baffle that reaches almost the bottom of the aeration basin. The mixed liquor is forced to flow from the bottom of this chamber to the surface, so that the hydrostatic pressure drop causes the air bubbles' release. The gentle hydraulic mixing in the chamber also contributes to the formation of large activated sludge flocs.

In both cases, the channel connecting the aeration basin with the secondary clarifier(s) should avoid highly turbulent flow conditions.

Figure 4.4 Hydraulically mixed degasification chamber

4.4 FEATURES OF SECONDARY CLARIFIER CONSTRUCTION

Construction details of secondary clarifiers are described in specialized literature on secondary clarifiers. Besides the IAWQ/IWA Scientific and Technical Report No. 6 (Ekama et al., 1997), much valuable information can be found on the web pages of design, engineering, manufacturing and construction companies involved in this business by using a general Internet search engine with keywords such as 'secondary clarifiers' and 'activated sludge separation'. The web pages of national and international professional organizations in the field of wastewater treatment such as EWA (European Water Association, www.ewa.de), IWA (International Water Association, www.iwahq.org.uk), WEF (Water Environment Federation, www.wef.org) and specialized web pages on activated sludge (www.scitrav.com/wwater/) can also be useful.

This chapter only describes those construction features of secondary clarifiers that are important in handling poorly-settling activated sludge.

4.4.1 Inlet structure with a flocculation zone

4.4.1.1 Principles of a flocculation zone

In the past the inlet structure of secondary clarifiers was designed similarly to those in primary clarifiers. The main purpose of this structure (called *the feed well*) was to dissipate the energy of the incoming mixed liquor and distribute evenly the flow throughout the cross-section of the clarifier. Usually the retention time in the feed well was not long enough for reflocculation of small activated sludge flocs. Current construction of secondary clarifiers typically provides a longer time for flocculation. This can be achieved in rectangular clarifiers by an additional compartment between the feed well and the sedimentation zone. In circular clarifiers the center feed well can be increased in diameter or an additional concentric flocculation zone provided (see Figure 4.5).

Flocculation process principles and design parameters are described together with construction recommendations for both circular and rectangular clarifiers in the specialized literature (e.g. Albertson (1992) and IWA Scientific and Technical Report No. 6 (Ekama *et al.*, 1997).

Figure 4.5 Flocculation zone with tangential flow into the feed well

4.4.1.2 Conventional flocculation zones with tangential flow regime

Flocculation zones in circular clarifiers are usually rather shallow so that the retention time required for flocculation is achieved by increasing the diameter of the zone. Mixing is provided by the tangential horizontal flow regime (Figure 4.6).

This arrangement has a tendency to allow a thick layer of scum (caused by hydrophobic filamentous microorganisms, fats and other floating materials) to accumulate on the flocculation zone (Figure 4.7). If the scum is not removed regularly it may putrefy and produce unpleasant odors.

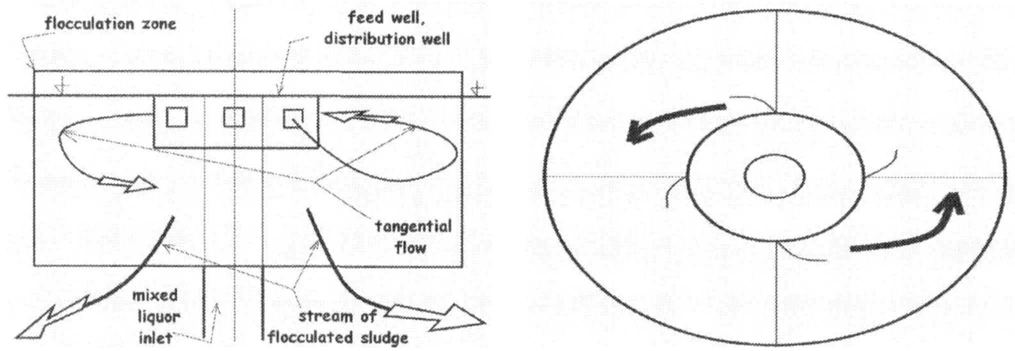

a) b)

Figure 4.6 Schematic view of flocculation zone with a predominantly tangential flow regime. a) vertical cross-section, b) horizontal cross-section.

Figure 4.7 Scum layer trapped in the ring of the flocculation zone

4.4.1.3 Novel flocculation zone with deflectors and a vertical flow regime

This major drawback of the conventional flocculation zone can be eliminated by changing the flow direction of the stream entering it. (Figure 4.8). The incoming mixed liquor is directed by the upper deflector towards the water surface so that it prevents the formation of a scum layer. This type of flocculation zone is much deeper and has a smaller diameter than a flocculation zone with a horizontal inlet flow.

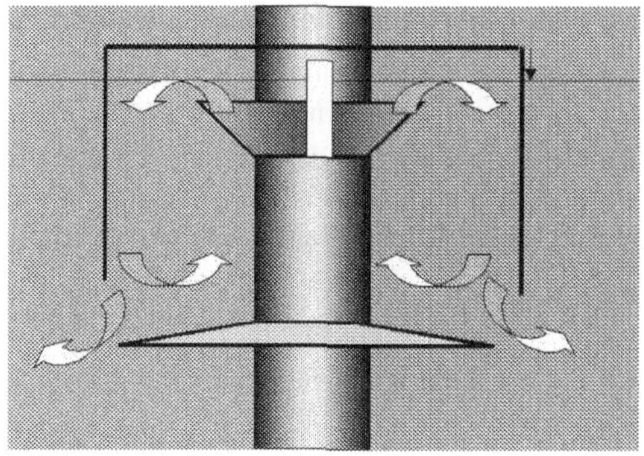

Figure 4.8 Flocculation zone with a vertical inlet flow (courtesy by KUNST, spol. s r.o., Hranice, Czech Republic)

4.4.2 Outlet structure

4.4.2.1 Peripheral vs. internal effluent launders

The main function of the outlet structure is to produce high quality final effluent containing low concentrations of suspended solids. One source of effluent suspended solids is from density currents. Usually the density currents are formed by differences in the densities of the mixed liquor and the sludge blanket at the bottom of the clarifier. These density currents can extend to the peripheral wall of the clarifier and turn the current towards the water surface. There are two scenarios in which the density current can reach the water surface:

4.4.2.1.1 Circular clarifier, peripheral effluent launder (with inboard weir)

Peripheral effluent launders are preferred by some clarifier designers because they allow the entire surface of the clarifier to participate in sedimentation. However this type of effluent launder is very susceptible to the effects of density currents. Albertson (1992) described how to deflect density currents back to the separation zone of the clarifier using a peripheral deflector (Stamford baffle) as depicted in Figure 4.9.

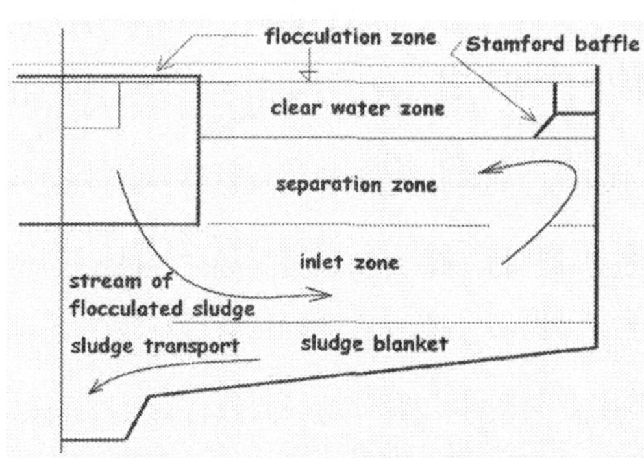

Figure 4.9 Schematic description of the effect of a Stamford baffle in a circular clarifier with a peripheral effluent launder

4.4.2.1.2 Circular clarifier, internal effluent launder (with inset double-sided weirs)

Another approach to the design of clarifier involves installing the effluent launders on a structure placed about 1/3 of the clarifier diameter from the outside wall. With this weir placement, a deflector (Stamford baffle) is not necessary because the density current turns back towards the separation zone away from weirs. In some designs two sets of internal launders have been used to increase effluent weir length. An example of a double-sided weir is shown in Figure 4.10.

Figure 4.10 Example of secondary clarifier with internal effluent launder with double-sided weir

4.4.2.1.3 Effluent launders in rectangular clarifiers with longitudinal flow

Figure 4.11 shows the flow pattern through a rectangular clarifier with longitudinal flow. To prevent density currents affecting effluent suspended solids levels, the effluent launders should be oriented parallel to the flow.

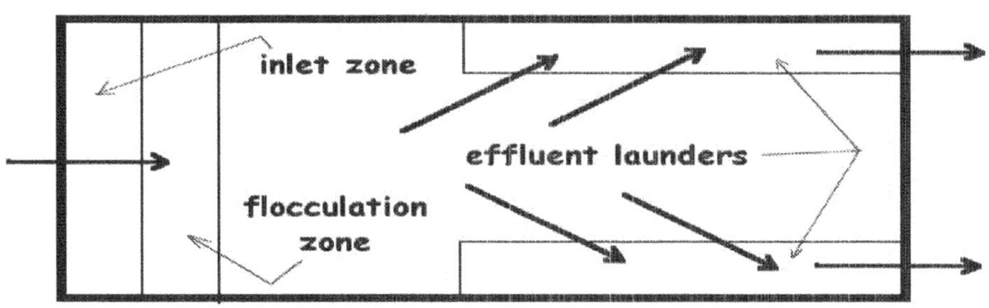

Figure 4.11 Scheme of a hydraulic flow through rectangular clarifier with a longitudinal flow

4.4.3 Scum baffles

The role of scum baffles is to protect the final effluent from the escape of floating biomass. As explained in Chapter 3, the floating biomass in a secondary clarifier results from:

- denitrification in the sludge blanket at the bottom of the clarifier
- flotation of foam formed by hydrophobic filamentous microorganisms and other trapped biomass

When the effluent weirs are not protected, the floating biomass escapes to some degree into the final effluent and deteriorates its quality. This can be prevented by inserting a scum baffle in front of the effluent weir (Figure 4.12 and 4.13). The installation of scum baffles also requires the installation of a skimming device to prevent the captured floating material from accumulating on the surface of the secondary clarifier.

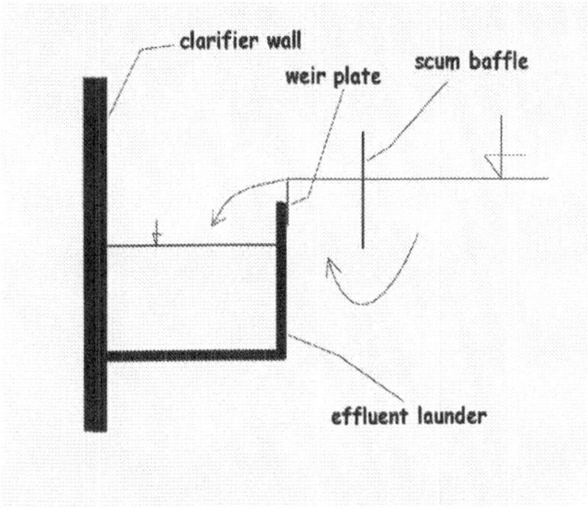

Figure 4.12 Effluent launder protected by scum baffle

Figure 4.13 The importance of scum baffle

4.5 EFFICIENT SCUM REMOVAL FROM THE SURFACE OF SECONDARY CLARIFIERS

4.5.1 Conventional scum boxes

The removal of floating biomass from the surface of circular secondary clarifiers is one of the most difficult tasks for the clarifier designer. Conventional scum removal devices such as those used in primary circular clarifiers (Fig. 4.14 and 4.15) are not suitable, because:

- the skimmer usually pushes the scum around the surface of the clarifier
- the surface area of the scum removal box is usually much smaller than the area of floating scum so that during one rotation of the travelling bridge only a small amount of the floating material is removed from the clarifier surface
- biological foam caused by filamentous microorganisms is much denser and more viscous than the "common scum" on primary clarifiers so that the scum box often becomes clogged.

a)

b)

Figure 4.14 a), b) Conventional skimmer and scum box – the area of scum box is too small to capture all of the floating material

a)

b)

Figure 4.15 a), b) Scum boxes clogged by viscous biological foam

4.5.2 "Travelling" scum boxes

Substantial improvement in scum removal efficiency can be achieved by installing the scum box on the travelling bridge together with the skimmer. In this way, the scum is skimmed to the box as the bridge rotates rather than only once each rotation. The captured scum is continuously pumped out of the box (Fig. 4.16).

a)

b)

Figure 4.16 a), b) Examples of a "travelling" scum box

4.5.3 Pneumatic systems

These systems take advantage of the fact that floating material is much lighter than water and can be moved by small movements of the surface. This movement is generated by pressurized air blown from the center of the clarifier over the water surface (Fig. 4.17). The air creates small waves which push the scum towards the peripheral scum launder.

The same effect can be achieved by installing a simple ventilator on the traveling bridge, which pushes the floating biomass along the skimmer to the scum box (Figure 4.18).

Figure 4.17 Pneumatic collection of scum in a circular secondary clarifier

Figure 4.18 Additional installation of a ventilator in front of the skimmer

4.6 REMOVAL OF SETTLED AND THICKENED SLUDGE

4.6.1 Effect on the final effluent quality

There are two traditional lines of thought concerning the amount of thickened activated sludge that should be allowed to accumulate in a secondary clarifier:

i) the amount of sludge in the bottom layer should be minimized because:
- a thick sludge layer allows denitrification leading to sludge rising
- in EBPR plants, phosphate release may occur in the thick sludge layer

ii) the sludge blanket may play a positive role because:
- it exhibits a filtration effect, especially when the flocculated stream is introduced into the sludge blanket (see Figure 4.9). This allows unflocculated microflocs to be "captured" in the settled sludge
- it represents an additional inventory of sludge which can be used to control F/M
- denitrification can contribute significantly to overall nitrogen removal efficiency

Modern secondary clarifiers (both circular and rectangular) are constructed with side wall depths of approximately 4 m. This allows operation with some sludge blanket. A procedure for the estimation of optimum sludge blanket thickness is presented in the ATV-DVWK Standard ATV A131 (2000).

Secondary clarifier operation with an optimized sludge blanket depth is important with respect to new legislation in many (European) countries which require the treatment of some wet weather flows. Storage of sludge in the clarifier can be used to prevent sludge wash-out during storm flows. The sludge removal system must have adequate capacity to prevent an excessive accumulation of the sludge in the secondary clarifier because this can lead to problems such as sludge rising and phosphate release.

4.6.2 Mechanical scrapers

Mechanical scrapers are used in both circular and rectangular clarifiers. The pricipal role of this device is not to "scrape" the sludge (i.e., to push it mechanically towards the hopper) but to keep it in a liquid stage. Secondary clarifiers with mechanical scrapers have sloped floors so that the main sludge transport mechanism is gravity flow. However, activated sludge is a biological material with a lot of extracellular biopolymers. Because it contains biopolymers, sludge is highly thixotropic so that the settled sludge quickly looses its ability to flow unless it is mechanically mixed at regular intervals. Because the major role of the scraper is to provide the energy necessary to overcome the thixotropic behavior of the sludge, the following design features are important:

- circular clarifiers: scrapers are mounted on a traveling bridge over the whole diameter of the clarifier; in this way the sludge blanket is mechanically disturbed twice in each rotation of the bridge
- rectangular clarifiers: scrapers are mounted on an endless chain which provides almost continuous movement of the sludge blanket.

4.6.3 Vacuum sludge removal

Vacuum sludge removal is used widely in the USA but not so widely in Europe where, if it is used, it is usually in rectangular clarifiers. The advantage of vacuum removal is that the thickened sludge is removed from the place where it settled rather than having to flow to a central sludge collection well or hopper. For this reason the bottom of the clarifier can be flat. The advantage of simpler construction is offset by the more complicated operation of the vacuum suction system. Sludge removal should be as frequent as possible because the thixotropic nature of the sludge will make its vacuum removal difficult. Initially the vacuum is created by a vacuum pump but subsequently the vacuum is created by the difference in water surface levels between the clarifier and the mixed liquor channel (Figure 4.19).

Figure 4.19 Rectangular secondary clarifier with the suction pipe system for vacuum removal of settled and thickened sludge

4.7 REFERENCES

Albertson O.E. (1992). Clarifier Design. In *Design and Retrofit of Wastewater Treatment Plants for Biological Nutrient Removal*. Randall C.W., Barnard J.L., Stensel H. Eds.,185-254. *TECHNOMIC Publ. Co.* Inc. Lancaster (PA) USA.

ATV-DVWK Standard A 131E (2000). *Dimensioning of Single-Stage Activated Sludge Plants*. GFA Publisher Hennef Germany.

Chudoba J., Dohanyos M., Wanner J. (1991). *Biological Wastewater Treatment* (In Czech). SNTL (State Publishing Company of Technical Literature) Prague.

Ekama G.A. et al. (1997). *Secondary Settling Tanks: Theory Modelling Design and Operation* Scientific and Technical Report No. 6 IAWQ/IWA London UK. 1-216.

Konicek Z., Burdych J. (1988). Effect of activated sludge process on secondary sedimentation tank efficiencies. *Water Sci & Technol*, **20**(4-5), 153-163.

5

Bulking and Foaming Control Methods

D. Dionisi, M. Majone and R. Ramadori

5.1 INTRODUCTION

In this chapter both control strategies for bulking in existing plants and design strategies for preventing bulking in new plants are discussed. The chapter deals mostly with filamentous bulking, which is the most common and most studied activated sludge separation problem. Control strategies for non-filamentous ("viscous bulking") are described in section 5.5.

Traditionally bulking studies have been conducted more systematically, associating identified filaments to specific experiences either in pure and mixed cultures, bench, pilot and full scale. The knowledge on bulking is wider than that on foaming, especially considering the minor information available on many foam microbial components, whose diversity appears quite large (Soddell, 1999): the main point is that there are much less experiences available on correctly identified foam microbial components and precise correlation with causes that caused their growth. For this reason this chapter will deal mainly on "Bulking Cure Methods", remaining the chlorination and mechanical removal the still more applied method to control foaming phenomena (Jenkins et al. 2004).

Filamentous bulking control methods are usually divided into specific and non-specific methods.

Specific methods are aimed at the removing the cause for filament proliferation, and are therefore targeted to a specific micro-organism or group of micro-organisms, whereas non-specific methods are aimed at reducing filament levels without any effect on the causes of their growth. Therefore they are independent of filament type.

The use of specific control methods is usually the more desirable method, because only when the cause of bulking is addressed can the plant be operated without any filament

 Activated Sludge Separation Problems: Theory, control measures and practical experience edited by V. Tandoi, D. Jenkins And J. Wanner. ISBN: 1900222841. Published by IWA Publishing, London, UK.

proliferation. While non-specific methods are temporary they are useful when the cause of the filamentous bulking cannot be determined immediately and a rapid resolution of the bulking problem is needed.

In this chapter each cause of bulking is described together with the corresponding specific control methods. At the end of the chapter non-specific control methods are described.

5.2 SPECIFIC CONTROL METHODS

Although the causes of filaments proliferation in activated sludge plants have been studied for many years, precise conclusions about the factors which favour filament growth cannot be drawn at present. Table 5.1 summarizes the present status of knowledge, of the various factors which have been reported as major causes of bulking. For each of these causes, the usually-adopted specific method for bulking control is also reported.

Table 5.1. Summary of causes of filamentous bulking, related filaments and suggested cure methods (Jenkins et al 2004)

Cause of filamentous bulking	Most common filaments	Suggested cure method
Low F/M	*H. hydrossys, Nocardia sp.*, *Thiotrix spp.*, Types 0041, 0675, 0092, 0803, 0914, 1851	Use of plants with substrate gradient (SBR, plug-flow systems or selectors)
Low dissolved oxygen	Type 1701, *S. Natans, H. hydrossys*	Increase dissolved oxygen level in the plant
Nutrient deficiency	*Thiotrix spp.*, type 021N, Nostocoida limicola II	Addition of the lacking nutrient
Fatty acids in wastewater, low F/M, low temperatures	*Michrotrix parvicella*	Use of anoxic or anaerobic selectors

5.2.1 Bulking due to low F/M

In activated sludge processes characterised by low (0,05-0,10 kg COD/kg VSS d) F/M ratio the concentration of COD in the aerobic tank is very low, specially in the case of completely mixed tank reactor (Figure 5.1)

According to the largely accepted kinetic selection theory developed by Chudoba and co-workers (1973), at these low substrate concentrations the growth of filaments is favoured with respect to floc-forming bacteria that are, on the other hand, favoured when substrate concentration is high. A different theory for explaining the selection for filaments at low substrate concentration is the "diffusion-based" selection (Martins *et al.*, 2003). According to this theory, floc-forming microorganisms face lower substrate concentration than filaments due to diffusion limitations inside the flocs. At low bulk substrate concentration this would lead to a lower growth rate of floc-formers than of filaments.

Bulking associated with low substrate concentration is linked to good activated sludge performance because low F/M values are associated with higher COD removals. In order to solve (or avoid) the problem without affecting substrate removal in the plant, the most useful strategy is to create a substrate concentration gradient inside the aeration tank reactor (i.e. the presence of zones whith different substrate concentrations). This goal can be achieved by modifying the activated sludge process configuration (shown in Fig. 5.1) by one of three methods (Fig. 5.2):

- continuous plug flow reactors (CPFR)
- completely mixed reactors with a contact zone placed ahead the main reactor (CMR)
- sequencing batch reactors (SBR).

In all three methods the aim of the process configuration is to achieve the removal of most of the substrate when it is present at high concentrations, where floc-formers are kinetically favoured, and to remove only a small fraction of the substrate at the low concentrations where filament growth is favoured. In SBR systems the desired substrate gradient is achieved in time, while in plug-flow and in selector systems the substrate gradient is achieved in space.

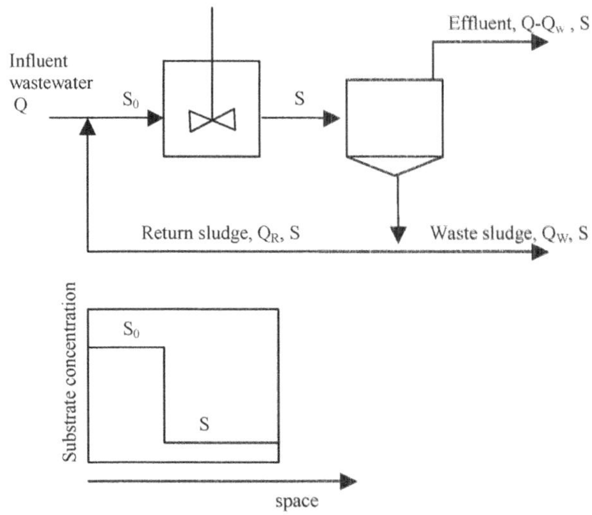

Figure 5.1. Conventional activated sludge process with continuously mixed aeration tank and substrate profile inside the aeration tank

5.2.1.1 Continuous plug-flow reactors

In this type of reactor a substrate profile is created along the reactor. The substrate gradient causes selection of floc-formers over filaments: floc formers are favoured in the first zones of the reactor, near to the inlet, where substrate concentration is high, whilst filaments are favored at the reactor outlet, where substrate is present at low concentration. Because most of the substrate is removed when its concentration is high, the overall result of this reactor is a kinetic selection that favours floc-formers. However large scale "corridor-type" reactors, which are often used in activated sludge plants and often considered as plug-flow reactors, usually have a significant longitudinal mixing due to the presence of aeration in aerobic zones, and mechanical mixing in anoxic or anaerobic zones. (Wanner, 1994). Therefore they are often not effective for bulking control: best results are obtained with a compartmentalization of the corridor type basin. In this case the system can be considered and designed as a selector.

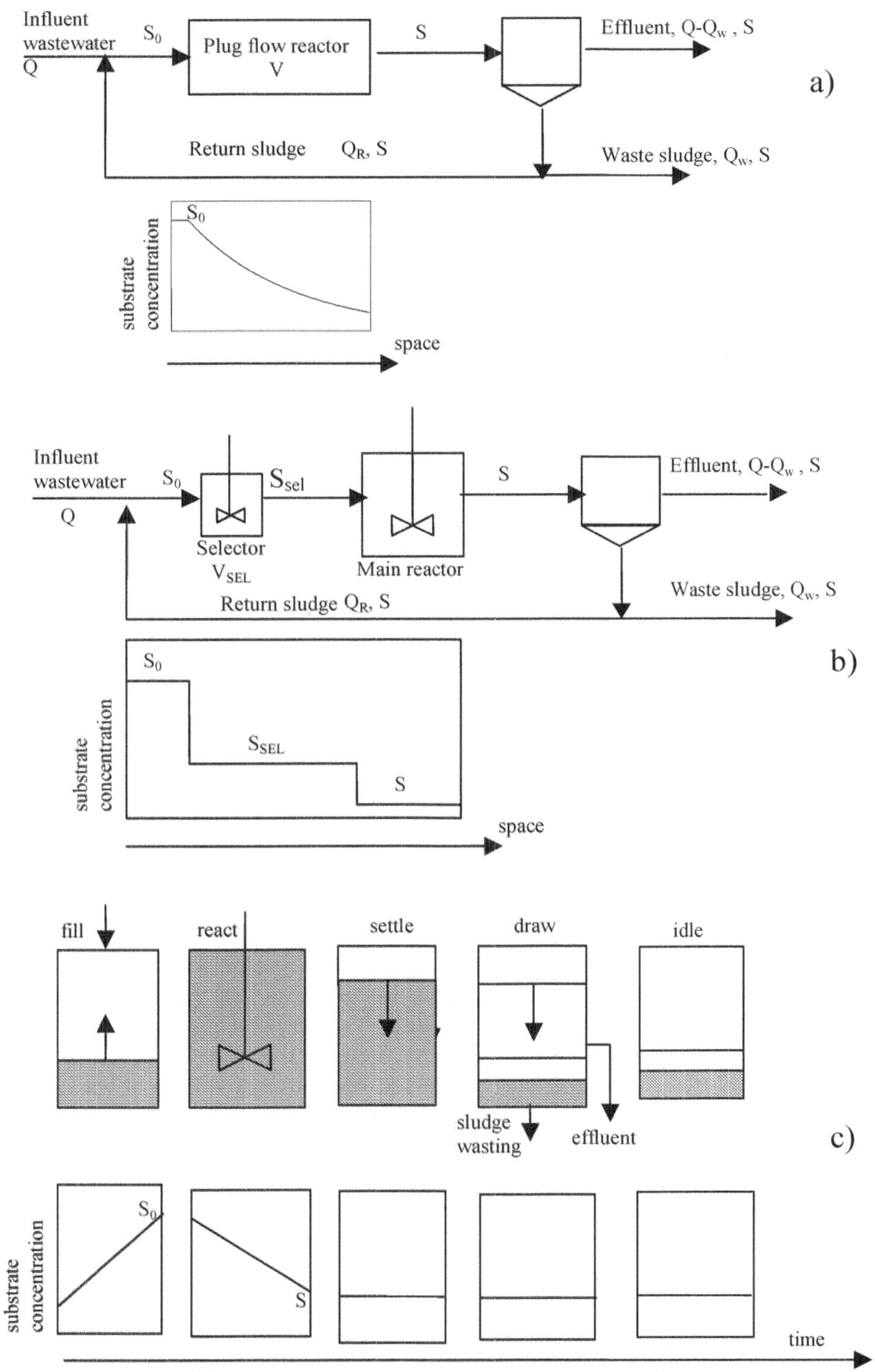

Figure 5.2 Activated sludge process flowsheets with a substrate gradient (a) plug- flow reactor, (b) selector, (c) sequencing batch reactor

5.2.1.2 Selectors

This is the most common scheme to combat filamentous bulking and the one which most closely resembles the original selector idea of Chudoba and co-workers (1973). The term selector usually refers simply to a smaller reactor placed ahead the main biological reactor. In the selector biomass removes COD, but due to the small selector volume and thus the substrate-biomass low contact time, COD concentration does not reach the low levels that favour filament growth. In the main biological reactor a low COD concentration is present but the limited amount of COD removed does not allow enough a filamentous organisms growth to cause bulking.

Proper selector design is critical. Too large a volume can cause a too high COD removal in the selector and thus lead to a COD concentration below the critical value that favours filament growth. On the contrary too small a volume could cause too much substrate removal in the main reactor, where could occur filamentous organism proliferation.

Selectors can be designed using either biokinetic models or empirical methods. In general empirical criteria are preferred, because of the uncertainty in kinetic parameters for the biokinetic models. Examples of both types of design criteria will be shown.

a) Design based on a biokinetic models

Design criteria proposed by van Niekerk *et al.* (1988). The design criteria were based on a mathematical description of the selector-aerator system in which filament and floc-former growth rates were described by the Monod relationship. A higher floc-former growth rate at high substrate concentrations and higher filament growth rate at low substrate concentrations was used. The problem of determining the optimum range of selector residence times which allows a higher growth rate for floc-formers than for the filaments was solved by setting up materials balances for the selector, the reactor and the settling tank and solving the resulting equations.

Another mathematical model of floc-former/filament competition which can be used for selector design purposes was developed by Kappeler and Gujer (1994a and b). This model takes into account the different fractions of the influent COD, and the inoculation of floc-formers and filaments from the sewer system.

As in the model described previously, the Kappeler and Gujer model allows for a selector design which minimizes the amount of filaments in the main reactor. Knowing the kinetic and stoichiometric parameters, simulations can be made for various selector volumes with a range of several design parameters such as sludge age, filament inoculation etc. Figure 5.3 shows the results of model calculation for optimum selector size. It is evident that in the typical ranges of activated sludge operating parameters the optimum volume of the selector is 5-10 % of the overall volume of the system (oxidation tank + selector).

As already pointed out, the main drawback of this design methodology is the large amount of kinetic and stoichiometric parameters that need to be known.

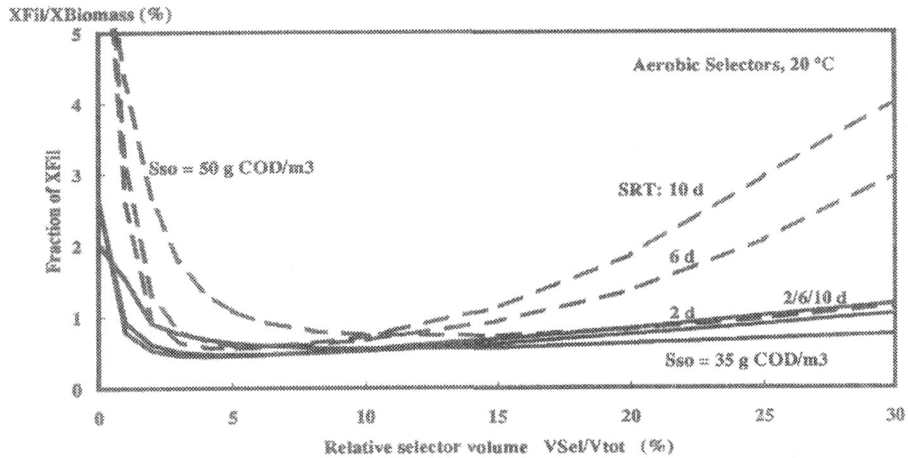

Figure 5.3 Calculations of the optimum selector size with the model by Kappeler and Gujer (1994b)

b) Design based on empirical criteria

Empirical selector design criteria have been proposed by several workers. Most useful are parameters that specify the selector hydraulic retention time (t or contact time) and the selector loading (B_x or activated sludge loading in the selector), defined as follows (see Figure 5.2):

Selector hydraulic retention time:

$$t = \frac{Q + Q_R}{V_{SEL}}$$

Selector loading:

$$B_X = \frac{Q \cdot S_0}{V_{SEL} X}$$

Table 5.2 shows literature values for these design parameters. Once the value of t or B_x has been chosen, the selector volume can be calculated.

c) Enhancing the selector effect: anoxic, anaerobic or compartmentalized selectors

The selector effect, can be enhanced if it can be operated under anoxic or anaerobic conditions. The improvement in selector perfomance usually obtained under anoxic or anaerobic conditions is due to the fact that with few exceptions the filaments responsible for bulking (and listed in Table 5.1) cannot use electron acceptors other than oxygen. The exceptions are some filaments, the so-called "all zone growers" (Wanner and Grau, 1989), that can grow under anoxic and anaerobic conditions. Examples of these are *M. parvicella* and Type 0092.

The implementation of anoxic or anaerobic selectors has some limitations: to use an anoxic selector nitrate must be available through nitrification and, to achieve rapid COD removal in anaerobic selectors, EBPR must be established. The suggested design criteria for anoxic selectors are reported in Table 5.3.

Table 5.2. Proposed empirical design criteria for aerobic selectors according to a) hydraulic retention time, b) selector loading.

a)

t (min)	Reference
15	Daigger *et al.*, (1985)
10-12	ATV (1989)
10-15	Eikelboom (1991)
15	Pujol (1992)
>10	Rensink and Donker (1992)
10-30	Prendl and Kroiss (1998)
11	Guida *et al.* (2001)
10-13	Martins *et al.* (2003)

b)

B_X (kg(BOD_5 or COD)/kgSS d)	Reference
>20 (COD)	Lee *et al.* (1982)
3 (BOD), 20 (COD)	ATV (1989)
>3	Chudoba and Wanner (1989)
5-6 (BOD)	Linne *et al.* (1989)
3.2-4.9 (BOD)	Daigger and Nicholson (1990)
3 (BOD)	Albertson (1991)
2-5 (BOD)	Eikelboom (1991)
4.2 (COD)	Still *et al.* (1996)
1.8-2.7 (BOD)	Nakhla and Lugowski (2003)

Table 5.3 Reported criteria for anoxic selector design

Parameter	Suggested value	Reference
Contact time	10-30 min	Hsu and Wilson (1992)
Activated sludge loading in the selector	2-6 kg COD/kg MLSS d	Wanner (1993)
Activated sludge loading in the selector	0.8-1.2 kg BOD_5/kg MLSS d	Marten and Daigger (1997)
Contact time	17-22 min	Andreasen *et al.* (1999)

For anaerobic selectors it is generally believed that the same HRT that is used in EBPR system can be used for the design of anaerobic selectors. Bortone *et al.* (1995) obtained good results with an anaerobic selector designed for an HRT of 45 min and Chevakidagarn *et al.* (2001) showed good results with an anaerobic selector HRT of 55-65 min. A similar anaerobic HRT (60 min) was shown to be effective in a study of Lee and Oleszkiewicz (2004).

The selector effect can be enhanced if the selector volume is compartmentalized: in this way the substrate gradient along the system is enhanced. As an example of design criteria for multiple compartment selectors, Jenkins *et al.* (1993) propose the following loading values for the three compartment selector:

- First compartment: 12 kg COD/(kg MLSS d)
- Second compartment: 6 kg COD/(kg MLSS d)
- Third compartment: 3 kg COD/(kg MLSS d)

Recently, Duine and Kunst (2001) found that growth of Type 021N and Type 0961 in an industrial WWTP could only be avoided with a three compartment aerobic selector with an HRT of 5-8 minutes per compartment. A corresponding completely-mixed selector was not successful in contrasting these filaments.

5.2.1.3 Sequencing batch reactors

Sequencing batch reactor (SBR) technology is becoming more and more widespread in the field of wastewater treatment (Wilderer *et al.*, 2001). The operation of a typical SBR is shown in Figure 5.2c. The typical sequence of phases is feed, reaction (oxic or anoxic), settle, effluent and sludge withdrawal and idle. SBRs offer many advantages over continuous flow systems, especially the possibility to perform all the treatment processes in a single tank.

The main operating parameters specific to SBR processes which are related to filamentous bulking control are the number of the cycles per day and the length of the feed period.

At a fixed organic load rate, the number of cycles per day determines the organic load rate per cycle, i.e., the extent of the temporal substrate gradients which are necessary for bulking control. By increasing the number of cycles per day, the organic load rate per cycle decreases and so the substrate gradients imposed to the biomass. Therefore, for obtaining a proper selection in favour of floc-forming microorganisms the number of cycles per day should not be too high. Usually, positive effects of SBRs on bulking control are reported with 4 or 6 cycles per day (Dionisi *et al.*, 2001; Martins *et al.*, 2003).

As far as the feed period is concerned, SBRs can select floc formers over filaments if the feed period is short enough or if feeding occurs under non-aerated conditions. With a too long aerated feed period the substrate will be removed simultaneously with feeding and the temporal substrate gradient necessary to avoid filament proliferation will not be achieved. A study by Martins *et al.* (2003), showed that, in a lab scale acetate-fed reactor, a feed period length of less than 15 min (6.2% of the total HRT) was needed to select for floc-formers and to obtain a low SVI: with a feed period length 15 min. or more (8.2% or more of the HRT) of *Sphaerotilus natans* proliferated and a high SVI was obtained.

5.2.1.4 Role of storage phenomena in microbial competition in substrate gradient processes

The positive effect of selectors, plug-flow configurations and SBRs on bulking control is explained, by Chudoba and co-workers (1973) by the different relationship between growth rates and substrate concentration for the two populations, or, by Martins *et al.* (2003), by the role of diffusion in the flocs. A different model for explaining the effect of systems with substrate gradients on population dynamics is based on storage phenomena. Storage polymers are intracellular reserve substances which are synthesized when the external substrate is in excess of that needed for cellular synthesis and are consumed when the external substrate is exhausted. The most common storage polymers are the polyhydroxyalkanoates (PHA), which are usually synthesised when volatile fatty acids are the substrate, and glycogen, when carbohydrates are the substrate. Storage usually occurs when normal cellular synthesis (i.e. synthesis of RNA, proteins and DNA) is prevented or slowed by some growth limitation: these growth limitations can be due to the lack of some macronutrient, as nitrogen or phosphorus or even sulfur (in this case the limitation is called external) or to the lack of the full metabolic apparatus (such as enzyme and RNA levels) needed to grow at maximum rate (in this case the limitation is called internal). In activated sludge plants treating domestic wastewaters external limitations are unlikely. Internal limitation can occur because of the very low growth rates and the frequently changing environment in which biomass is grown (Majone *et al.*, 1999). The more dynamic the conditions the more likely that storage phenomena will occur. Thus in plants with selectors or plug-flow aeration regime tanks the first mechanism of substrate uptake is usually storage.

The relevance of storage as the first substrate uptake mechanism in activated sludge plants has been recently recognised from a modelling point of view (Gujer *et al.,* 1999).

The role of storage in microbial competition relies on the fact that various populations seem to have different storage capabilities and rates (Wanner, 1994). Storage ability seems to be widespread among floc-formers, whereas only few filamentous species are able to store substrate. As a consequence, the substrate gradients (with space or time) encourage storage phenomena favouring floc-formers over filaments.

5.2.2 Bulking due to low nutrient concentration

The term "nutrient" in activated sludge systems usually refers chemical elements, other than carbon, oxygen and hydrogen, that are essential for biomass synthesis. Nutrients needed in major amounts are nitrogen (N) and (P) phosphorus: the minimun COD:N:P ratio in the influent stream is usually assumed to be 100:5:1. Table 5.4 shows the usual nutrient requirements of activated sludge. Domestic wastewaters are normally rich in nutrients (N and P often have to be removed by specific removal processes), while industrial wastewaters, can be lacking in one or both of these nutrients.

Table 5.4 Nutrient requirements of activated sludge process (Grau, 1991)

Nutrient	Requirement (g/kg BOD removed)
N	50
P	10
Fe	12
Ca	6.2
K	4.5
Mg	2.0
Mo	0.43
Zn	0.16
Cu	0.15
Co	0.13
Na	0.05

Data reported by Wagner *et al.* (1982) and Richard *et al.* (1985) show that filament growth is enhanced with respect to floc-former growth when N and P levels are low. The study by Richard *et al.* (1985), shows that the filamentous bacterium Type 021N, has a high N uptake rate when grows under N- deficient conditions. A study performed by Simpson *et al.* (1991) relates filaments proliferation to the lack of metals: SVI in the studied plants quickly decreased after solutions of metals salts were added to the plant. In one case the filaments that were successfully controlled by metal addition were identified as *Microthrix parvicella* and Type 021N.

When bulking can be definitely related to nutrient deficiency, the control method is to add the deficient nutrient and it is easily implemented. A successful case study is reported by Switzenbaum *et al.* (1992) where the proliferation of *Nostocoida limicola* II due to P deficiency in the plant was controlled by increasing the sludge age so that the rate of nutrient utilization by the biomass decreased and P was no longer limiting.

5.2.3 Bulking due to low dissolved oxygen concentrations

One of the most frequent causes of filamentous bulking is low dissolved oxygen (DO) concentration in the aeration tank.

A model for the effect of DO on microbial competition can be set-up by assuming that DO is the only limiting substrate and that biomass growth rate is dependent on DO concentration according to a Monod-like kinetic equation. Experimental data on pure cultures of Type 1701 and *Sphaerotilus natans* show that these filamentous bacteria have values of the Monod half-saturation constants for DO in the range 0.01-0.03mg/L, whereas typical values for pure cultures of floc-formers are about 0.07 mg/L (Jenkins et al 2004). It is clear that, with DO concentrations found in activated sludge plants (usually approx. 2 mg/L) both populations can grow at their maximum rate and therefore DO concentration should not play a role in population dynamics. However the role of DO is relevant because the half-satauration constants values given above are intrinsic values and do not take into account the diffusional resistances to DO diffusion within the flocs. Taking diffusional resistance into account the DO concentration inside the floc is much lower than in its external part and thus growth of floc-formers inside of the floc can be DO-limited, whereas growth of filaments which protrude from the floc in the bulk liquid is not. With the Monod-like kinetic equations describing the effect of DO, this phenomenon can be considered by introducing "apparent" half-saturation constants, whose values are higher than the intrinsic values for both populations, but much higher for the floc-formers than for the filaments.

The relevance of diffusional resistance in determining the DO profile inside the floc is dependent on the amount of substrate removed by the floc, which is in turn dependent on the organic loading. A correlation (Table 5.5) between organic loading minimum DO concentration needed to avoid filamentous bulking was developed by Palm *et al,* 1980. Values of DO concentrations required to avoid filament proliferation can be calculated and can be provided for proper aerator design.

Table 5.5 Minimum dissolved oxygen values necessary to avoid filamentous bulking as a function of organic load rate (Palm *et al.*, 1980)

Organic load rate (kg COD/kg MLVSS d)	Dissolved oxygen (mg/L)
0.3	1
0.5	2
0.75	3
0.9	4

5.2.4 Bulking due to fatty acids in the influent stream: control methods for *Microthrix parvicella*

Microthrix parvicella is often reported in activated sludge plants and its growth is particularly difficult to prevent. Most of the studies performed with pure cultures of this micro-organism show that its growth is enhanced by low temperatures and by the presence of long chain fatty acids in the influent. Pure culture studies have shown that this micro-organism is only able to grow under aerobic or microaerofilic conditions, while it cannot grow under anoxic or anaerobic conditions (Tandoi *et al.*, 1998). It recently has been reported (Nielsen *et al.*, 2002) that this micro-organism can take up and store long chain fatty acids under anaerobic conditions and this ability should explain its proliferance in P-removal plants.

The most frequently used technique for controlling *Microthrix parvicella* has been the use of selectors. Eikelboom reported that aerobic selectors are not effective (Eikelboom, 1994), but good results were obtained with a plug-flow aerobic or an anoxic/aerobic reactor (Mamais *et al.*, 1998). The best results were obtained using anoxic and anaerobic reactors in series (Kruit *et al.*, 2002) or anoxic selectors alone (Davoli *et al.*, 2002). In the first case the growth of *Microthrix parvicella* was avoided with a process including an anaerobic reactor (contact time 80 min) followed by an anoxic selector (contact time 30 min) or with a process including a four-compartment anaerobic selector (total contact time 35 min) followed by a two-compartment anaerobic reactor (contact time 55 min). Both of the processes described above perform full nutrient removal and are based on a sequence of anoxic and aerobic reactors. The experiences described by Davoli *et al.* (2002) showed that *Microthrix parvicella's* growth could be controlled by anoxic selectors with contact times of 25-69 min. Recently the use of aluminium polichiloride (PAX 14) las been shown effective in controlling *M. parvicella* (see section 5.3.3)

5.2.5 Microbial and enzymatic preparations

The addition of microbial and enzymatic preparations to control filamentous bulking has been much more limited than the use of biocides and flocculating agents. Inamori *et al.* (1991) reported that two ciliated protozoa were able to feed selectively on filaments like Type 021N or *Sphaerotilus natans*. Yaguchi *et al.*, 1991, showed that a lytic enzyme specifically active against Type 021N could be produced by soil microorganisms.

Many commercial microbial and enzymatic preparations exist but their effectiveness has not been studied in a controlled systematic fashion; this does not mean that in individual specific cases the use of enzymes cannot be successful, especially for bulking or foaming by *M. parvicella* due to fatty acids in the influent.

5.3 NON SPECIFIC CONTROL METHODS

Often all of the factors leading to sludge bulking are unidentified, so that the use of non-specific control methods is widespread. The use of non-specific control methods can be the first method of controlling bulking, before a cause and effect relationship can be found and a specific control method implemented. Non specific control methods usually consist of the addition of chemicals. The chemicals which are most often used are:

- oxidising agents (chlorine, ozone, hydrogen peroxide)
- weighting and flocculating agents (salts of iron and aluminium, lime, polymers and talc)

Tables 5.6 and 5.7 report the amounts of the various chemicals that have been used.

Table 5.6 Reported criteria for oxidising agent dosage: a) chlorine; b) ozone; c) hydrogen peroxide

a)

Chlorine (g/kg SS d)	Reference
1-15	Jenkins *et al.*, 1982
5-15	Neethling *et al.*, 1985
0.43-1	Adey and McClintock, 1996
1.42-3.07	Saayman *et al.*, 1996
4-10	Marten and Daigger, 1997
6-8	Ramirez *et al.*, 2000
4-8	Ramirez *et al.*, 2001

b)

Ozone (g/kg SS d)	Reference
2	Van Leeuwen, 1992
12-48	Collignon *et al.*, 1993
6.7-16.6	Goi *et al.*, 1994
0.36-1.42	Saayman *et al.*, 1996
16.5-27.6	Kim and Somiya, 1998

c)

Hydrogen peroxide (g/kg SS d)	Reference
1.0	Vedry *et al.*, 1988
8.0	Van Leeuwen, 1992
1.5-9.5	Saayman *et al.*, 1996

Table 5.7 Reported criteria for weighting or flocculating agents dosage: a) Iron salts or lime; b) polymers; c) talc

a)

Substance	Amount	Reference
$FeSO_4$ $7H_20$	7 mg Fe/L	ATV (1989)
$FeSO_4$ $7H_20$	35 mg Fe/L	Wanner (1994)
$Fe_2(SO_4)_3$ (43-45 %)	7.5 mg Fe/L	Echeverria et al. (1993)
lime	10-15 mg Ca/kg SS d	Wanner (1994)

b)

Type of polymer (commercial name)	Amount	Reference
cationic	<1 mg/L	WEF Design Manual (1992)
cationic charged polyacrylamide	0.5 mg/L	Shao *et al.*, (1997)
cationic charged polyacrylamide, MW 10^7 (Zetag 88 N)	8 mg/L	Vanderhasselt and Verstraete, (1999)
cationic charged polymer, MW 6 10^6	3-12 mg/L	Seka *et al.*, (2001)
(Keystone KF 4249), MW 2.5-3.5 10^6	1 mL polymer/ 10 L wastewater	Der-Fong (2005)

Table 5.7 (cont'd)

c)

Substance (Commercial name)	Amount	Reference
Talc (PE8418)	0.6-1 kg/kg SS	Eikelboom and Grovenstein (1998)
Multi-component additive: 1% biocide, 98.5% talc, 0.5% polymer	0.35-1 kg/kg SS	Seka *et al.*, (2001), Seka *et al.* (2003)
Talc (Aquatal)	0.6-1 kg/kg SS	Bideault et al., (1997)
Talc (Aquatal)	0.7 kg/kg SS	Clauss *et al.*, (1999)
Talc (Aquatal)	0.5-8.5 g/L	Vanderhasselt and Verstraete (1999)

5.3.1 Oxidising agents

The aim of oxidising agent (biocide) addition is to kill filamentous organisms without affecting floc-formers. This is possible because filaments protrude from the floc and are therefore more subject to the toxic agent. Because of its potentially deteriorative effects on floc-formers, the correct choice of addition point and of the amount of the toxicant added should be carefully chosen. According to the scheme of Figure 5.4, three main locations for biocides addition are possible:

- directly in the aeration tank;
- in a side stream recycle from the aeration tank;
- in the return sludge.

The best performance is usually obtained by adding the toxicant to the return sludge stream, because of the higher amount of solids which are exposed to it. Almost all of the data reported in Table 5.6 refer to addition of the toxicant to the return sludge. In cases where the return flow rate is too low, addition to the return stream is inappropriate and the best location for toxicant addition is to the aeration tank.

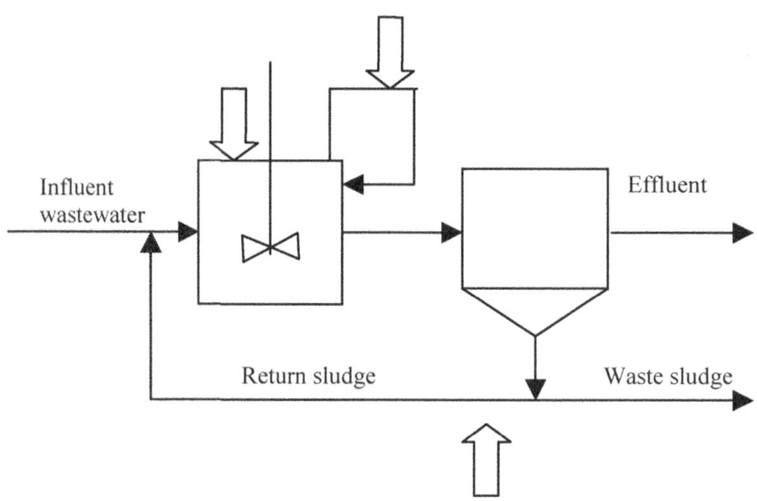

Figure 5.4 Possible locations (indicated by the arrows) for addition of oxidising agents

In addition to the daily dosage of the biocide reported in Table 5.6, other important parameters in the dosage strategy are the concentration of the biocide at the dose point (defined as the ratio of the mass of biocide dosed per day and the flow rate past dose point) and the frequency of exposure of activated sludge to chlorine dose (Wanner, 1994). Recommended values are, for the first parameter, 15-20 mg Cl_2/L in order to avoid floc-formers damage, and for the second one a value higher than 2.5-3 d^{-1}.

The most widely used biocide agent is Cl_2, provided as chlorine solution or sodium hypochlorite solution. It is also the oldest control measure against bulking (Smith and Purdy, 1936). Table 5.6 shows that the amount of chlorine fed to activated sludge systems varies in the range 1-15 g Cl_2/kgVSS d.

A recent study (Caravelly *et al.*, 2003) using pure cultures of *Sphaerotilus natans* and *Acinetobacter anitratus* showed that the amount of Cl_2 needed to obtain a more than 80% inhibition of respirometric activity of both micro-organisms was about 12 mg Cl_2/gVSS. In another study on a phosphorus removal sludge, Chang *et al.* (2004) reported a significant improvement of setlling properties (both SVI and zone settling velocity) of a sludge with high abundance of *Thiotrix* with chlorine dosages up to 40 mgNaOCl/gSS, but, on the other hand, also noticed a worsening of substrate uptake rate and phosphorus removal at chlorine dosages higher than 2.5 mgNaOCl/gSS.

In order to better choose the optimum chlorine dosage with the aim of killing filaments without affecting the floc-forming population, some control methods have been developed recently based on a commercial stain mixture (LIVE/DEAD BacLight, Ramirez *et al.,* 2000) that allows living and dead microorganisms to be evaluated microscopically. In this way the effect of chlorine on the biocenosis can be monitored easily and chlorine can be added on the basis of filament and floc damage. Alternatively, chlorination effects can be assessed by traditional microscopic observation: in particular the presence of small flocs and the small amounts of microfauna are frequently linked to chlorine overdosage. Information about the amount of chlorine that can be added without affecting nutrient removal can be found in Saayman *et al.* (1996, see below).

Microthrix parvicella, seems to be particularly difficult to control even with chlorine dosage. Hwang and Tanaka (1998) reported, that amounts of chlorine up to 20 g Cl_2/kgSS had no effect on *Microthrix parvicella* filaments, whilst this amount was highly detrimental to floc-forming micro-organisms. An amount of 200 g Cl_2/kgSS was needed to break down more than half of the coils of the filament. As the disinfecting effect of chlorine is thought to be a function of penetration rate through the cell wall (White, 1982), the poor success in *Microthrix parvicella* bulking control with chlorine dosage could be due to the hydrophobic cell wall of this filament, which prevents the biocide agent from penetrating into the cytoplasm.

The addition of chlorine to activated sludge systems has caused some concerns about possible effluent quality deterioration due to the formation of halogenated organic compounds. Van Leeuwen and Van Rossum (1990) showed in batch tests that the chlorination of activated sludge lead to the formation of halogenated organics such as $CHCl_3$ and $CHBrCl_2$ but in pilot continuous flow activated sludge systems these compounds were never detected in the final effluent due to air stripping.

The first reports of the use of hydrogen peroxide (H_2O_2) are dated more than 30 years ago (Caropreso *et al.*, 1974) and its use has been limited by its high cost (van Leeuwen, 1992). It has the advantage of not leaving toxic residual since it is ultimately converted to water and oxygen:

$$2\ H_2O_2 \rightarrow 2H_2O + O_2$$

Experiences with ozonation are the more recent. Its dosage is more costly than chlorination (van Leeuwen, 1988; van Leeuwen, 1992): ozone has to be generated on site and requires costly equipment. Like H_2O_2, its addition does not leave toxic residuals.

A comparative cost analysis of the three biocides has been reported by van Leeuwen (1992).

The effects of the three biocides on biological nutrient removal was investigated by Saayman *et al.* (1996). They found that a chlorine dosage higher than 1.4 g Cl_2/kgSS d had a detrimental effect on P removal, whereas nitrification and denitrification were virtually unaffected up to 3 g Cl_2/kgSS d. H_2O_2 had no effects on nitrification and denitrification at all the tested dosages 1.5-9.5 g H_2O_2/kgSS d. It had a small (and temporary) inhibitory effect on P removal at the maximum dosage. O_3 addition (0.36-1.4 g O_3/kgSS d) had no negative effect on biological nutrient removal (indeed, P removal was sometime enhanced).

5.3.2 Weighting or flocculating agents

Bulking can be controlled without the addition of potentially toxic oxidizing agents by adding weighting or flocculating agents.

Traditionally-used flocculating agents are chemicals such as lime or Fe and Al salts. These compounds are also used in the chemical P precipitation and overcome the bridging or diffuse floc structure associated with high filament levels, yielding an immediate improvement in settling. As a further beneficial effect, the addition of iron salts can precipitate sulphides and thus depress the growth of sulphide utilising filamentous microorganisms as *Thiotrix* or *Beggiatoa* (Echeverria *et al.*, 2002). As alternative flocculating agents, polymers (usually cationic) can be used.

Divalent cations such as Ca^{2+} and Mg^{2+} can also be used to favor the formation of biological flocs. According to the divalent cation bridging theory (DCBT) (Sobeck and Higgins, 2002), divalent cations bridge the negatively charged functional groups which are present on exocellular biopolymers and therefore promote good floc formation. When the ratio of monovalent-divalent cations in the wastewater is high (more than approximately 2, expressing the ions as meq/L), divalent cations are displaced from the biomass by monovalent cations and settling properties (SVI and effluent solids) deteriorate. Therefore in order to control bulking, the ratio of monovalent-divalent cations in mixed liquor should be lowered to values less than 2, either by increasing the divalent cation concentrations or by decreasing the monovalent cation concentrations. Successful case studies have been reported. (Higgins *et al.*, 2004a and b).

Talc-based additives have also been proven successful (Bidault *et al.*, 1997, Claus *et al.*, 1999). Talc is a hydrated magnesium silicate often combined with other minerals. Talc is insoluble, it is not toxic to bacteria and it is compatible with other chemicals commonly used in wastewater treatment plants. For the purpose of floc weighting, talc minerals are usually used as water suspensions and are added directly into the aeration tank. The action of talc is to modify flocculation in spite of the filaments by creating larger flocs. These larger flocs usually entrap the filaments and prevent them from decreasing sludge settleability. Experimental studies show that the average floc dimensions and the settling velocities increase with higher talc additions. It is evident that, when talc-based additives are added the amount of TSS and ash in the system increase, but the total mass of sludge after dewatering on the filter-press remains unchanged, because more water is expelled from excess sludge due to the better dewatering properties of the sludge to which the talc was added (Bideault *et al.*, 1997).

The positive effect on settling properties of biocides and of weighting agents can be combined in one single multi-component additive (Seka *et al.*, 2001, 2003): it has been shown that an additive in powder form composed of (w/w) 1 % biocide (a quaternary ammonium compound), 98.5 % talc and 0.5 % polyelectrolytes could simultaneously both weight the flocs (due to the action of the polymer and of talc) and reduce the filaments (due to the action of the biocide). Through the addition of the multi-component additive, both immediate and long-term effects (for several months) were obtained.

5.3.3 New findings: use of alluminum salts to control *Microthrix parvicella*

Recently it has been proposed that the addition of the polyalluminum chloride interferes strongly with the growth of the filament *Microthrix parvicella* by causing either a flocculating or specific toxic effect (Eikelboom, 1997). This approach was successfully applied in several full-scale plants at a dosage of from 1.5 to 4.5 g Al^{3+}/kg MLSS, depending on SRT (Roels *et al.* 2002, Badoer *et al.* 2003), This strategy is promising and investigation on its basic mechanisms should be conducted.

5.4 CONTROL OF NON-FILAMENTOUS BULKING ("VISCOUS BULKING")

The problem of non-filamentous (viscous) bulking has not been studied as much as filamentous bulking. However, two main reasons seems to be evident for this type of bulking:

a) a high amount of fatty acids and oleic compounds in the wastewater;
b) an insufficient amount of certain nutrients, such as nitrogen and phosphorus.

Based on this evidence, specific control methods for non-filamentous bulking can be adopted.

In a laboratory scale study (Novak *et al.*, 1993), the decrease in the oleic acid proportion in the influent wastewater (from 33 to 19 % of the total influent COD) was supposed to be the reason for a shift from bad-settling to well-settling sludge in about 3 weeks. In an SBR fed with brewing wastewater (Peng *et al.*, 2003), non-filamentous bulking was observed due to nitrogen limitation (influent BOD:N ratio 100:0.94) whereas a well-settling sludge was obtained when the influent composition was more balanced (BOD:N 100:4). At the Detroit Wastewater Treatment Plant (Franklin *et al.*, 2001) the problem of non-filamentous bulking, which was apparently due to low concentrations of soluble phosphate (<0.2mg/L), was corrected by increasing the soluble phosphate feed to the plant.

As far as non-specific control methods are concerned, they seem usually ineffective in the control of non-filamentous bulking. However, a successful case study was reported (van Leeuwen *et al.*, 1989) where an O_3 dosage of 1 g/(kg MLSS d) decreased the SVI from 125 to 70mL/g, mainly by preventing excessive Zoogloeal growth.

5.5 AVOIDING POOR SETTLING PROPERTIES: ALTERNATIVE SEPARATIONS OF ACTIVATED SLUDGE

The activated sludge separation problems described in this STR are all related to low settling rates which cause the secondary sedimentation problems. One control that solves the problem at its origin is to use a different device for the sludge-supernatant separation: of the available alternative methods the only one which has been used with some success is membranes (Stephenson *et al.,* 2000). The scheme of a bioreactor equipped with a membrane for solid-liquid separation (a membrane broreactor, MBR) is shown in figure 5.5. The use of membranes avoids the effect of filamentous micro-organism on settling because the filterability of the sludge seems to be independent of the number of filaments present. The solids concentration which can be handled by an MBR is much higher than that which can be handled by a secondary clarifier. The biomass concentrations that can be achieved in an MBR activated sludge process are much higher than in conventional systems (up to 40 g/L). This, in addition to the advantages of not suffering from bulking problems, MBR activated sludge also has the advantage of a smaller footprint.

The disadvantage of MBRs is their high cost. Membrane costs are approximately proportional to plant size, while in conventional plants there is an economy of scale. Additional disadvantages are membrane fouling, which requires periodic cleaning, and high aeration requirements for preventing membrane fouling.

However, also in MBRs, as in conventional activated sludge processes with secondary sedimentation, the operating conditions of the activated sludge plant are important in determining the feasibility of the membrane separation process (Liao *et al.*, 2004). As an example it was observed that the sustainable flux across membranes increased by increasing the sludge age from 3-8 to 33 d, likely because of a lower membrane fouling due to the lower concentration of extracellular polymeric substances (EPS) at high sludge age (Chang and Lee, 1998). Membrane fouling was also lower at low organic loading, because of the lower EPS content (Nagaoka *et al.*, 2000).

Therefore, in an hypothetical future where membranes will replace secondary clarifiers, we cannot expect to no longer deal with activated sludge separation problems. Simply, activated sludge processes will have to be designed in order to select for sludges with "good filterability properties", instead of "good settling properties".

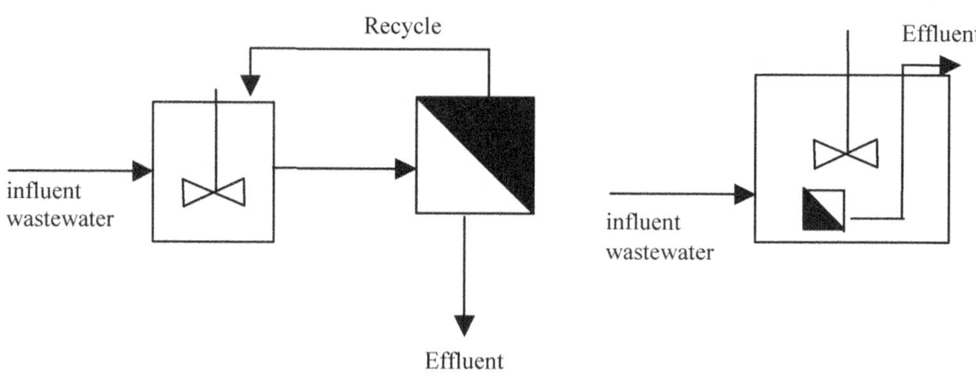

Figure 5.5 Alternative separation for activated sludge: membrane bioreactors. Sidestream (left) and submerged (right). Source: Stephenson *et al.*, 2000

5.6 REFERENCES

Adey A.H., Mcclintock S.A. (1996). Control of filamentous bulking in a complete-mix activated sludge process treating high-strength industrial wastewater. *Proc. - WEFTEC '96 Annu. Conf. Expo. 69th*, 3, 395-404.

Albertson O.E. (1991). Bulking sludge control – Progress practice and problems. *Water Sci & Technol* **23**(4-5), 835-846.

Andreadsen K., Agertved J., Petersen J.O., Skaarup H. (1997). Improvement of sludge settleability in activated sludge plants treating effluents from pulp and paper industries. *Water Sci & Technol* **40**(11-12), 215-221.

ATV Working Group 2.6.1 (1989). Prevention and control of bulking sludge and scum. *Korrespondenz Abwasser*, **36**(13) 165-177.

Badoer S., Cuomo M., Miana P., Fabris M., Rossetti S., Tandoi V. (2003). Bological scum by *Microthrix parvicella* in a large wastewater treatment plant in Northern Italy: causes and solutions. IWA Regional Conference "Water key to sustainable development in Africa", Cape Town 14-17 September.

Bidault A., Clauss F., Helaine D., Balavoine C. (1997). Floc agglomeration and structuration by a specific talc mineral composition. *Water Sci & Technol*, **36**(4) 57-68.

Bortone G., Cech J.S., Bianchi R., Tilche A. (1995). Effects of an anaerobic zone in a textile wastewater treatment plant. *Water Sci & Technol*, **32**(9-10), 133-140.

Caravelli A.H., Contreras E.M., Giannuzzi L., Zaritzky N.E. (2003). Modeling of chlorine effect on floc-forming and filamentous microorgnaisms of activated sludges. *Wat Res*, 37, 2097-2105.

Caropreso F.E., Raleigh C.W., Brown J.C. (1974). Attack bulking sludge with H_2O_2 a microscope. *Bull California Water Poll Control Assoc*, **44**.

Chang I.S., Lee C.H. (1998). Membrane filtration characteristics in membrane coupled activated sludge systems- effect of physiological states of activated sludge on membrane fouling. *Desalination*, **120**, 221-233.

Chang W.C., Jou S.J., Chien C.C., He J.A. (2004). Effect of chlorination bulking control on water quality and phosphate release/uptake in an anaerobic-oxic activated sludge system. *Water Sci & Technol*, **50**(8), 177-183.

Chevakidagarn P., Orth H., Ratanachai C. (2001). The use of a selector when improving the efficiency of the conventional activated sludge process under tropical temperature conditions. Poster 48 *Proc. of the 3rd IWA Specialised Conference on Microorganisms in Activated Sludge and Biofilm Processes*. Rome 13-15 June, Tandoi V., Passino R., Blundo M.C. Eds.

Chudoba J., Grau P., Ottova V. (1973). Control of activated sludge filamentous bulking-II Selection of microorganisms by means of a selector. *Wat Res*, **7**, 1389-1406.

Chudoba J., Wanner J. (1989). Notes on oxic anoxic and anaerobic selectors. *Newsletters of the IAWPRC Specialist group on activated sludge population dynamics*, **1**(2), 46-49.

Clauss F., Helaine D., Balavoine C., Bidault A. (1998). Improving activated sludge floc structure and aggregation for enhanced settling and thickening performances. *Water Sci & Technol*, **38**(8-9), 35-44.

Collignon A., Martin G., Martin N., Laplanche A. (1993). Treatment of bulking by ozonation. Mechanisms of ozone on microorganisms. *Trib. Eau* **46**, 46-57.

Daigger G.T., Robbins M.H., Marshall B.R. (1985). The design of a selector to control low F/M filamentous bulking. *J Water Pollut Control Fed*, **16**, 35-53.

Daigger G.T., Nicholson G.A. (1990). Performance of four full-scale nitrifying wastewater treatment plants incorporating selectors. *Res J Water Poll Control Fed*, **62**, 676-683.

Davoli D., Madoni P., Gugliemi L., Pergetti M. (2001). Testing the effect of selectors in the control of bulking and foaming in small activated-sludge plants. *Water Sci & Technol*, **46**(1-2), 495-498.

Der-Fong J. (2005). Effects of synthetic polymer on the filamentous bacteria in activated sludge. *Biores Technol*, **96**(1), 31-40.

Dionisi D., Majone M., Tandoi V., Beccari M. (2001). Sequencing batch reactor: influence of periodic operation on physiological state microbial composition and performance of activated sludge in biological wastewater treatment. *Ind Eng Chem Res*, **40**, 5110-5119.

Duine A., Kunst S. (2001). Control of bulking sludge caused by Type 021N and Type 0961 in an industrial wastewater treatment plant with an aerobic selector. *Water Sci & Technol*, **46**(1-2), 29-33.

Echeverria E., Seco A., Ferrer J. (1993). Control of activated sludge settleability using preaeration and preprecipitation. *Water Res*, **27**, 293-296.

Eilkelboom D.H. (1991). The role of competition between floc-forming and filamentous bacteria in bulking activated sludge. *Biological approach to sewage treatment process: current status and perspectives.* Madoni P. Ed., Perugia Italy, Luigi Bazzucchi Center, 143-149.

Eilkelboom D.H. (1994). The *Microthrix parvicella* puzzle. *Water Sci & Technol,* **29**(7), 271-279.

Eilkelboom D.H. (1997). Control of *Microthrix parvicella* by addition of PAX 14. TNO Report R 97/305.

Eikelboom D.H., Grovenstein J. (1998). Control of bulking in a full scale plant by addition of talc (PE 8418). *Water Sci & Technol,* **37**(4-5), 297-301.

Franklin C., Hanson C., Verma K., Packman J. (2001). Case study; viscous bulking-importance of nutrient monitoring. *Conference Proceedings - Water Env Fed Annual Conference and Exposition 74th* Atlanta, USA, 13-17 Oct, 4794-4822.

Goi M., Nishimura T., Kuribayashi S., Okouchi T., Murakami T. (1994). An experimental study to suppress scum formation accompanying the abnormal growth of *Nocardia* by adding ozone in the aeration tank. *Water Sci & Technol,* **30**(11), 231-234.

Gousailles M., Castex P. (1988). Reduction of filamentous sludge bulking in municipal sewage treatment by hydrogen peroxide *Eau Ind Nuisances,* **122**, 99-100.

Grau P. (1991). Criteria for nutrient-balanced operation of activated sludge process. *Water Sci & Technol,* **24**(3-4), 251-258.

Guida M., Cesaro G., Lipardi I.L., Melluso G. (2001). A full scale application in the control of the filamentous bulking generated by Type 021N and *Thiotrix sp. Water Sci & Technol,* **46**(1-2), 507-510.

Gujer W., Henze M., Mino T., van Loosdrecht M.C.M. (1999). Activated sludge model No 3. *Water Sci &Technol,* **39**(1) 183-193.

Higgins M.J., Tom L.A. Sobeck D.C. (2004a). Case study I: application of the divalent cation bridging theory to improve biofloc properties and industrial acytivated sludge system perfomance-direct addition of divalent cations. *Water Engineering Resources,* **76**, 344-352.

Higgins M.J., Sobeck D.C., Owens S.J., Szabo L.M. (2004b). Case study II: application of the divalent cation bridging theory to improve biofloc properties and industrial acytivated sludge system perfomance-using alternatives to sodium-based chemicals. *Water Engineering Resources,* **76**, 353-359.

Hsu M., Wilson T.E. (1992). Activated sludge treatment of municipal wastewater- USA practice. In *Activated Sludge Process Design: Theory and Practice* (Eckenfelder W.W., Grau P. Eds.), 37-68, Technomic Publishing Co., Inc. Lancaster PA

Hwang Y., Tanaka T. (1998). Control of *Microthrix parvicella* foaming in activated sludge. *Water Res* **32**, 1678-1686.

Inamori Y., Kuniyasu Y., Sudo R., Koga M. (1991). Control of the growth of filamentous microorganisms using predacious ciliated protozoa. *Water Sci & Technol,* **23**(4-6), 963-971.

Jenkins D., Richard M.G., Daigger G.T. (2004). *Manual on the Causes and Control of Activated Sludge Bulking and Foaming and other solids separation problems,* 3rd Edn. Lewis Publishers, New York.

Kappeler J., Gujer W. (1994a). Development of a mathematical model for aerobic bulking. *Water Res,* **28**, 303-310.

Kappeler J., Gujer W. (1994b). Verification and application of a mathematical model for "aerobic bulking". *Water Res,* **28**, 311-322.

Kim Y.R., Somiya I. (1998). Effect of ozonization on filamentous bulking control of activated sludge, *Environ Conser Eng,* **27**, 590-597.

Kruit J., Hulsbeek J., Visser A. (2002). Bulking sludge solved?! *Water Sci & Technol,* **46**(1-2), 457-464.

Lee Y., Oleszkiewicz J.A. (2004). Bench-scale assessment of the effectiveness of an anaerobic selector in controlling filementous bulking. *Envir Technol,* **25**, 751-755.

Lee S.E., Koopman B.L., Jenkins D., Lewis R.F. (1982). The effect of aeration basin configuration on activated sludge bulking at low organic loading. *Water Sci & Technol,* **14**(6-7), 407-427.

Liao B.Q., Bagley D.M., Kraemer H.E., Leppard G.G., Liss S.N. (2004). A review of biofouling and its control in membrane separation bioreactors. *Water Env Res,* **76**, 425-436.

Linne S.R., Chiesa S.C., Rieth M.G., Polta R.C. (1989). The impact of selector operation on activated sludge settleablity and nitrification: pilot scale results. *J Water Poll Control Fed* ,**61**, 66-72.

Majone M., Dircks K., Beun J.J. (1999). Aerobic storage in the activated sludge processes. The state of the art. *Water Sci & Technol,* **39**(1), 49-60.

Mamais D., Andreadakis A., Noutsopoulos C., Kalergis C. (1998). Causes of and control strategies for *Microthrix parvicella* bulking and foaming in nutrient removal activated sludge systems. *Water Sci & Technol,* **37**(4-5), 9-17.

Marten W.L., Daigger G.T. (1997). Full scale evaluation of factors affecting the performance of anoxic selectors. *Water Engineering Resources*, **69**, 1272-1281.

Martins A.M.P., Heijen J.J., van Loosdrecht M.C.M. (2003). Effect of feeding patterns and storage on the sludge settleability under aerobic conditions. *Water Res*, **37**, 2555-2570.

Martins A.M.P., Pagilla K., Heijnen J.J., van Loosdrecht M.C.M. (2004). Filamentous bulking sludge-a critical review. *Water Res*, **38**, 793-817.

Nagaoka H., Kono S., Yamanishi S., Miya A. (2000). Influence of organic loading on membrane fouling in membrane separation activated sludge process. *Water Sci & Technol*, **41**(10-11), 355-362.

Nakhla G., Lugowski A. (2003). Control of filamentous organisms in food-processing wastewater by intermittent aeration and selectors. *J Chem Technol and Biotechnol*, **78**, 420-430.

Neethling J.B., Jenkins D., Johnson K.M. (1985). Chemistry microbiology and modelling of chlorination for activated sludge bulking control. *J Wat Poll Control Fed*, **57**, 882-889.

Nielsen P.H., Roslev P., Dueholm T.E., Lund Nielsen J. (2002). *Microthrix parvicella* a specialized lipid consumer in anaerobic-aerobic activated sludge plants. *Water Sci & Technol*, **46**(1-2), 81-90.

Novak L., Larrea L., Wanner J., Garcia-Heras J.L. (1993). Non-Filamentous activated sludge bulking in a laboratory scale system. *Water Res*, **27**, 1339-1346.

Palm J.C., Jenkins D., Parker D.S. (1980). Relationship between organic loading dissolved oxygen concentration and sludge settleability in the completely mixed activated sludge process. *J Water Pollut Control Fed*, **52**, 223-232.

Peng Y., Gao C., Wang S., Ozaki M., Tagikawa A. (2003). Non-filamentous sludge bulking caused by a deficiency of nitrogen in industrial wastewater treatment plant. *Water Sci & Technol*, **47**(11), 289-295.

Prendl L., Kroiss H. (1998). Bulking sludge prevention by an aerobic selector. *Water Sci & Technol*, **38**(8-9), 19-27

Pujol R. (1992). Control of filamentous bulking by the contact zone: design and results. *Newsletters of the IAWPRC Specialist Group on Activated Sludge Population Dynamics*, **4**(1), 24-26.

Ramirez G.W., Alonso J.L., Villanueva A., Guardino R., Basiero J.A., Bernecer I., Morenilla J.J. (2000). A rapid direct method for assessing chlorine effect on filamentous bacteria in activated sludge. *Water Res*, **34**, 3894-3898.

Ramirez G.W., Alonso J.L., Basiero J.A., Benecer I., Morenilla J.J. (2001). Survey of the filamentous populations using a rapid method for assessing chlorine effect in activated sludge. *Proceedings of the 3rd IWA specialised conference on Microorganisms in Activated Sludge and Biofilm Processes* Rome 13-15 June, paper 145. Tandoi V., Passino R., Blundo M.C. Eds..

Rensink J.H., Donker H.J.G.W. (1991). The effect of contact tank operation on bulking sludge and biosorption processes. *Water Sci & Technol*, **23**(4-6), 857-866.

Richard M.G., Shimizu G.P., Jenkins D. (1985). The growth physiology of the filamentous organism Type 021N and its significance to activated sludge bulking. *J Water Pollut Control Fed*, **57**, 1152-1162.

Roels T., Dauwe F., Van Damme S., De Wilde K., Roelandt F. (2002). The influence of PAX-14 on activated sludge systems and in particular on *Microthrix parvicella*. *Water Sci & Technol*, **46**(1-2), 487 – 490.

Saayman G.B., Schutte C.F., van Leeuwen J. (1996). The effect of chemical bulking control on biological nutrient removal in a full scale activated sludge plant. *Water Sci & Technol*, **34**(3-4), 275-282.

Seka A.M., van de Wiele T., Verstraete W. (2001). Feasibility of a multi-component additive for efficient control of activated sludge filamentous bulking. *Water Res*, **35**, 2995-3003.

Seka A.M., van de Wiele T., Verstraete W. (2003). Full-scale evaluation of a multi-component additive for efficient control of activated sludge filamentous bulking. *Envir Technol*, **23**, 67-72.

Shao Y.J., Jenkin s D. (1989). The use of anoxic selectors for the control of low F/M activated sludge bulking. *Water Sci & Technol*, **21**(6-7), 609-619.

Simpson J.R., List E., Dunbar J.M., (1991). Bulking sludge: a theory and successful case studies. *J Inst Water and Environ Manag*, **5**, 302-311.

Smith R.S., Purdy W.C. (1936). The use of chlorine for correction of sludge bulking in the activated sludge process. *Public Health Report*, **51**, 1-617.

Sobeck D.C., Higgins M.J. (2002). Examination of three theories of cation-induced bioflocculation. *Water Res*, **36**, 527-536.

Soddell J. (1999): Foaming, In: The Microbiology of activated sludge. Seviour R.J., Blackall L.L Eds., 1-422, Kluwer Academic, London.

Stephenson T., Judd S., Jefferson B., Brindle K. (2000). *Membrane Bioreactors for Wastewater Treatment*, IWA publishing London UK.

Still D.A., Ekama G.A., Wentzel M.C., Casey T.G., Marais CvR. (1996). Filamentous organism bulking in nutrient removal activated sludge systems. Paper 2: Stimulation of the selector effect under aerobic conditions, *Water SA*, **22**, 97-118.

Switzenbaum M.S., Plate T.R., Woodworth B.K.(1992). Filamentous bulking in Massachusetts: extent of the problem and case studies. *Water Sci & Techno,l* **25**(4-5) 265-271

Tandoi V Rossetti S Blackall LL Majone M (1998). Some physiological properties of an Italian isolate of *Microthrix parvicella Water Sci & Technol*, **37**(4-5), 1-8.

Vanderhasselt A., Verstraete W. (1999). Short term effects of additives on sludge sedimentation characteristics. *Water Res*, **33**, 381-390.

van Leeuwen J. (1988). Bulking control with ozonation in a nutrient removal activated sludge system. *Water SA*, **14**, 119-124.

van Leeuwen J. (1989). Ozonation for non-filamentous bulking control in an activated sludge plant treating fuel synthesys wastewater. *Water SA*, **15**, 127-132.

van Leeuwen J. (1992). A review of the potential application of non-specific activated sludge bulking control. *Water SA,* **18**, 101-106.

van Leeuwen J., van Rossum PG (1990). Trihalomethane formation during sludge bulking control with chlorine, *J Inst Water and Environ Manag*, **4**, 530-534.

van Niekerk A., Jenkins D., Richard M.G. (1988). A mathematical model of the carbon-limited growth of filamentous and flocforming organisms in low F/M sludge. *J Water Pollut Control Fed*, **60**, 100-106.

Wagner F. (1982). Study of the causes and prevention of sludge bulking in Germany *Bulking of Activated Sludge- Preventative and Remedial Methods* (Chambers B., Tomlinson E. J. Eds), 29-40, Chichester Ellis Horwood Ltd.

Wanner J., Grau P. (1989). Identification of filamentous microorganisms from activated sludge: a compromise between wishes needs and possibilities. *Water Res*, **23**, 883-891.

Wanner J. (1993). Kinetic and metabolic selection in controlling the filamentous organisms in activated sludge systems. *Prevention and control of Bulking Activated Sludge*. Jenkins D., Ramadori R., Cingolani L. Eds, Perugia Italy, Luigi Bazzucchi Center 17-42.

Wanner J. (1994a). Activated sludge bulking and foaming control. Technomic Publishing Co Inc, Lancaster USA

Wanner J (1994b). The implementation of bulking control in the design of activated sludge systems *Water Sci & Technol*, **29**(7), 193-202.

WEF Manual of Practice No 8 and ASCE Manual and Report on Engineering Practice No 76 (1992). "Population dynamics and control" in *Design of Municipal Wastewater Treatment Plants*, **1**, 538

White G.C. (1982). *Handbook of Chlorination*. Van Nostrand Reinhold, New York.

Wilderer P.A., Irvine R.L., Goronszy M.C. (2001). Sequencing batch reactor technology. *Scientific and Technical Report no 10*, IWA Publishing, London UK.

Yaguchi J., Chigusa K., Ohkubo Y. (1991). Isolation of microorganisms capable of lysing the filamentous bacterium "Type 021N" (Studies on lytic enzyme against the filamentous bacterium "Type 021N" Screening Studies). *Water Sci & Technol*, **23**(4-6), 955-962.

6

Experience in Various Countries

6.1 INTRODUCTION

This chapter reports experiences made in different countries to face bulking and foaming problems. This chapter is a tentative to give a whole picture of the approaches used worldwide: it can not be homogenous in style, content and descriptions, because in different countries different scientific approaches and experimental research were developed and carried out; consequently several control methods and solutions as well were adopted. Nevertheless, to produce a strategy to face the problems, an overview of the efforts made in 10 countries is the starting point, where to look for actions made, results obtained, and for suggestions.

6.2 AUSTRALIA (R. SEVIOUR, K.LINDREA, E. SEVIOUR, C. MCKENZIE)

6.2.1 General situation

Australia is the largest island continent in the world. Most of it is desert, and the vast majority of its people live in large cities like Melbourne, Sydney and Brisbane, along the eastern coastline. Thus, there exist enormous variations in, for example, their average rainfall, mean winter and summer seasonal temperatures and even dietary and cultural habits. All of these variables would be expected to affect the operational parameters of activated sludge plants, and hence their filamentous bacterial populations. This country does not lend itself readily to any generalisations.

Consequently, the survey data presented here must be interpreted cautiously. Often any operational data accompanying the biomass samples examined have been provided by plant personnel differing markedly in their understanding of the process, and with access to different (often no) amounts of analytical data. Thus, any cause and effect relationships mentioned here are at best tentative.

Furthermore, plants in Western Australia, South Australia, Northern Territory, and Tasmania have not, as far as we know, been extensively surveyed at all, and so nearly all the available data come from Victoria, New South Wales and Queensland. Even then in most of these surveys (Fig 6.1), only a single sample from each plant has been taken for analysis, providing information of limited, and only descriptive value. Our work shows that the populations of filamentous bacteria in a single plant can often vary considerably over a short period, for reasons unfortunately which are not usually clear.

However, work here and in Linda Blackall's laboratory at the University of Queensland has successfully led to the culture of many Australian filamentous bacteria (Blackall *et al* 1994;1996; Bradford *et al* 1996; Seviour *et al* 1997), and so their taxonomic positions have now been resolved for the first time. Furthermore, the availability of gene probes (Amann *et al* 1995) based on the 16S rRNA sequences of these filamentous bacteria now provide the means to determine unequivocally the biogeography and biodiversity of the filamentous bacteria globally. Thus one very important question to be addressed is whether the filamentous bacteria we 'identified' on the basis of their microscopic feature in Australia are the same as those sharing the same microscopic features seen in other countries. The same question can be addressed in a different way. As we now know that morphology is a very unreliable indicator of true relatedness, it is equally important to resolve whether a single morphotype (eg Type 1863 or Type 021N) is in fact a single bacterium, or an assemblage of quite different bacteria which all happen to look sufficiently similar under the microscope to make discrimination between them difficult. Australian studies have started to provide answers to these questions (eg Seviour *et al* 1997) for some filamentous bacteria, but it seems that these answers depend on the particular filament in question.

It is also clear from surveys carried out in Australia, and other countries, that many foams contain few or no filaments. Instead their microbial community is dominated by unicellular gram positive bacteria which we call 'Actinomycetes' (Seviour *et al* 1989). Many of the Mycolata (see later) which grow as filaments can fragment as part of their life cycle into single cells, and incidentally, would not be quantifiable using the microscopic methods described in the literature for quantifying filamentous bacteria (eg Jenkins *et al* 1993). We have successfully isolated these unicellular foam forms in pure culture, where they then grow as filaments typical of the Mycolata. One of these isolates has been identified as *Nocardia farcinica*, an opportunistic pathogen, and never previously reported in foams (Stratton *et al* 1996). Also, the acid fast stain suggests that many foams contain unicellular acid fast bacteria (probably *Mycobacterium* spp) which may also be undesirable pathogenic bacteria (Soddell *et al* 1999). So any international discussion on these troublesome bacteria should bear in mind the possibility that people talking about the same morphotype in different countries or from different plants, may in fact be discussing quite different bacteria. Therefore, any specific conclusions reached regarding control measures may not be generally relevant to all plants.

The other problem of international communication is in the names given to some of these bacteria. For example we now know that *Nostocoida limicola I, II and III* are in fact at least three totally unrelated bacteria, even though most survey data, including our own, have lumped them all together as *N. limicola*. The term 'Nocardia' is still commonly used to describe many foaming bacteria, even though we, and others (Soddell *et al* 1998;

Goodfellow *et al* 1996) have shown that organisms with the same typical branching morphology as *Nocardia amarae*, now named *Gordonia amarae* (eg Goodfellow *et al* 1994) belong to several different bacterial genera, with quite different physiologies. The term NALO (*Nocardia amarae*-like organisms) or more correctly GALO (*Gordonia amarae*-like organisms) should in our view be replaced by the generic term of Mycolata, to describe this group of bacteria, in preference to the commonly used term nocardioforms which is taxonomically less precise.

The other problem in comparing survey data from different countries and different groups is in how relative filament abundances are recorded ie dominant, excessive etc (Soddell 1999). The system used in our laboratory and throughout this chapter (i.e. number of categories) differs to some work performed elsewhere (i.e. Pujol *et al* 1991).

This brief review will attempt to summarise what is currently known about the distribution of these filamentous bacteria in Australian activated sludge plants, taking account of the difficulties in interpreting such data which were mentioned above. Several studies differing in the number and geographic spread of plants sampled have been performed in this country, and will be dealt with chronologically.

6.2.2 Foaming filamentous bacteria in Australian Activated Sludge Plants

Foams can easily be seen on the surfaces of the aeration tanks and so foaming is an easily recognised problem. Blackall (1987) seems to have been one of the first to survey foams in Australia. She looked at 51 plants in Queensland, 8 in NSW, 3 in Victoria and 2 in Western Australia. The microbiological data were restricted to the occurrence or otherwise in these plants of three morphotypes, *M. parvicella*, *Nocardia amarae* (now *Gordonia amarae*) and a novel branching Gram positive bacteria, she called Pine tree like organisms (PTLO), and later classified as a new species of *Nocardia*, *N. pinensis* (Blackall et al 1989). This organism, which will be discussed in more detail later, has since been placed into a new genus *Skermania* as *S. piniformis* (Chun et al 1997). Her data revealed that both *G. amarae-like* organisms and *S. piniformis* were commonly seen in foams in plants in Queensland with this problem, but *M. parvicella* was not. Foaming could not be correlated with any single design or operational feature of these plants like aeration system, DO levels, sludge age, which was not always available, or loading, although all non-foaming plants were under loaded. Some of the Mycolata were isolated using the technique of micro manipulation, and all the NALO (GALO) cultured (45% of plants) were *N. amarae* (*G. amarae*); *S. piniformis* was obtained in pure culture from 8 plants in Queensland, the only state Blackall (1986) recorded its presence. In a later, more extensive survey (Seviour et al 1989), involving systems of many configurations, both conventional and EBNR, 31 plants in NSW, 18 in Victoria and 17 in Queensland were examined. About half of the plants had foams when single samples were obtained. A fuller microscopic examination of these was undertaken and more filament Types 'identified'. However no distinction was made between *Nostocoida limicola* I, II or III (i.e. all referred to as *N. limicola*) or Eikelboom Types 0041 and 0675, which are conventionally separated on trichome diameter (Eikelboom & van Buijsen 1983), a character deemed too unreliable for this purpose (Seviour et al 1989). The data obtained (Fig. 6.1) demonstrate like the earlier survey that the major foaming bacteria in Australia are NALO (GALO) and *S. piniformis*, and also *M. parvicella* which occurred much more frequently than in the earlier survey. (Eikelboom) Type 0092 although frequently seen was never recorded as being the dominant filament. GALO and *S. pinensis* together with Type 1701 were commonest in Queensland, and Type 1701 was never detected in Victoria, while *M. parvicella* was much more common in Victoria than in Queensland. The reasons for such trends are unclear, but differences in climate and or diet (Eikelboom 1991) may be partly responsible.

Soddell & Seviour (1994) obtained data to suggest *S. piniformis* was in fact common in activated sludge plants in Queensland and also NSW and Victoria in both foams (26%) and mixed liquor (17.3%) samples (Fig. 6.1). They noticed considerable variations in the branching patterns of this bacterium in activated sludge plants (i.e. both branching angles and branching frequency), but isolates from 27 of these plants in pure culture were very similar in their phenotypic properties and all were considered to belong to the same single species (Soddell & Seviour 1998). This situation is of course quite different to the GALO, which we know all have a similar indistinguishable microscopic appearance, but contain organisms belonging to several different genera (Soddell 1999), and we suspect widely differing physiologies.

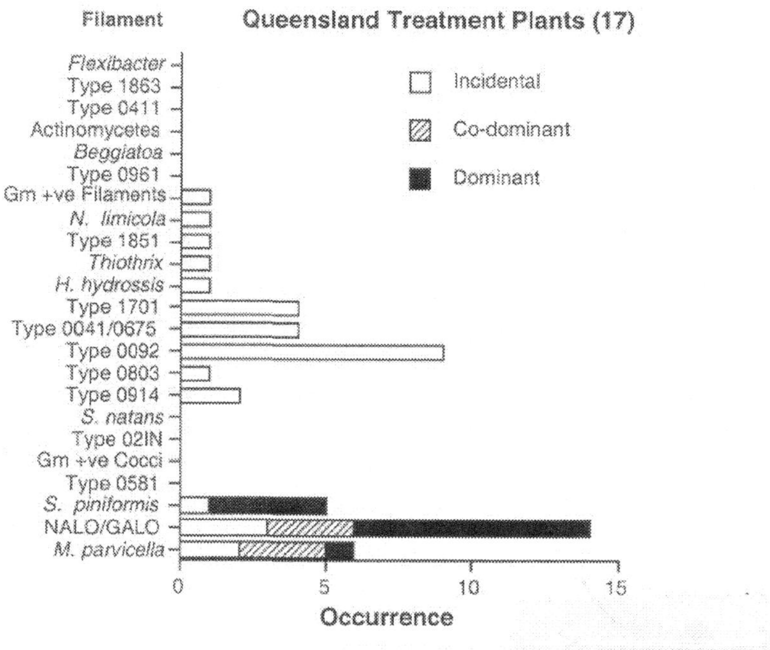

Fig. 6.1 Survey data of Seviour et al. (1994) showing major filamentous bacteria in bulking and foaming activated sludge plants in NSW, Victoria and Queensland

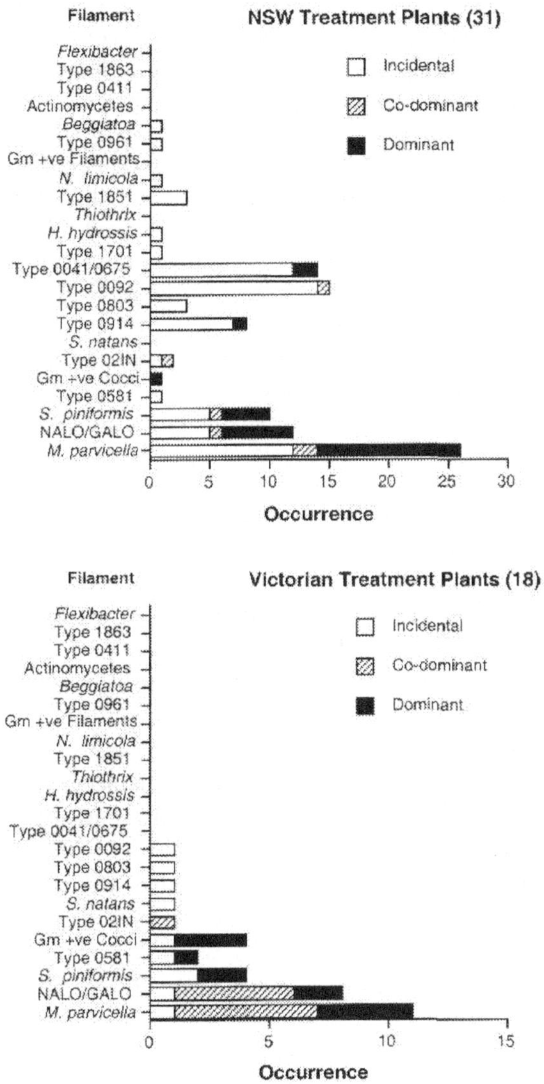

Fig. 6.1. (continued).

6.2.3 Foaming and Bulking Filamentous Bacteria in Activated Sludge Plants

Several surveys have also been performed where both mixed liquor and foam samples (if foaming) from a single plant have been examined, to attempt to elucidate which organisms may be true foaming bacteria and help understand the causes. One large survey was carried out by Seviour *et at* (1994) who examined single samples from 65 plants in Victoria and NSW (not Queensland), including some which had been sampled in the earlier survey. Of these, 53 bulked and 44 had a foaming incident at the time of sampling. However, what constitutes a bulking sludge is not as clear as is the problem of foaming, and much of the information on whether those plants bulked came from plant operators who were often uncertain. Many plants did not carry out settling tests with their mixed liquor, and so often the evidence was anecdotal. More than half of these plants (37) were bulking (based on DSVI or SVI values) and foaming, and only 2 plants showed neither symptom, illustrating how common these problems are in Australia. Where plants foamed, they tended to foam continuously (i.e. it was not a seasonal problem). Neither operational problem appeared to relate exclusively to any specific operational or design parameter, except possibly loading, since under loaded plants appeared to be more prone to both than overloaded systems. The main findings are that in these two states, *M. parvicella* was the dominant organism in both bulking and foam, samples. Type 0041/0675 was also very frequent and as expected from previous studies, GALO and *S. piniformis* were more commonly seen in foams than mixed liquor samples. Some of the Eikelboom filament morphotypes were successfully isolated into pure culture in this study, and in some cases (eg *M. parvicella*, Types 0803, 0411, 0092, 1863) these cultures provided the means whereby their true taxonomic positions could be determined for the first time using 16S rDNA sequence data (Blackall *et al* 1994; 1996; Bradford *et al* 1996; Seviour *et al* 1997).

A similar survey was reported by Bradford (1998) using single samples from plants (29) in Queensland, involving a total of 33 mixed liquor samples and 23 foam samples (Fig 6.2). Unlike all previous Australian surveys, Bradford (1998) used both microscopy and fluorescence *in situ* hybridisation (FISH) to identify her filamentous bacteria. She was able to demonstrate that FISH could detect some filaments hidden in the flocs eg *S. piniformis*, which could not be seen with a light microscope. Also FISH often gave different (correct?) identifications for some filaments 'identified' (wrongly) under the microscope, illustrating the perils inherent in this microscopic method. By selecting the appropriate probes FISH was able to suggest the presence in some samples of novel, previously uncharacterised GALO, again emphasising how little we know and understand about the diversity of foaming bacteria (Soddell *et al* 1998; Goodfellow *et al* 1996).

Unfortunately, probes are not yet available for all the filamentous bacteria seen in activated sludge, and so as yet, only an incomplete picture can be established. Some of the results she obtained were different to those in the earlier surveys, which may reflect a recent shift in the type of plant configuration now used. However, *M. parvicella* was still not often dominant in these Queensland plants (agreeing with Seviour *et al* 1989) and some plants which were foaming in the Seviour *et al* (1989) study, no longer did, and *vice versa*. *S. piniformis* could not be detected at all in non-foaming plants, disagreeing with the earlier study of Soddell & Seviour (1994).

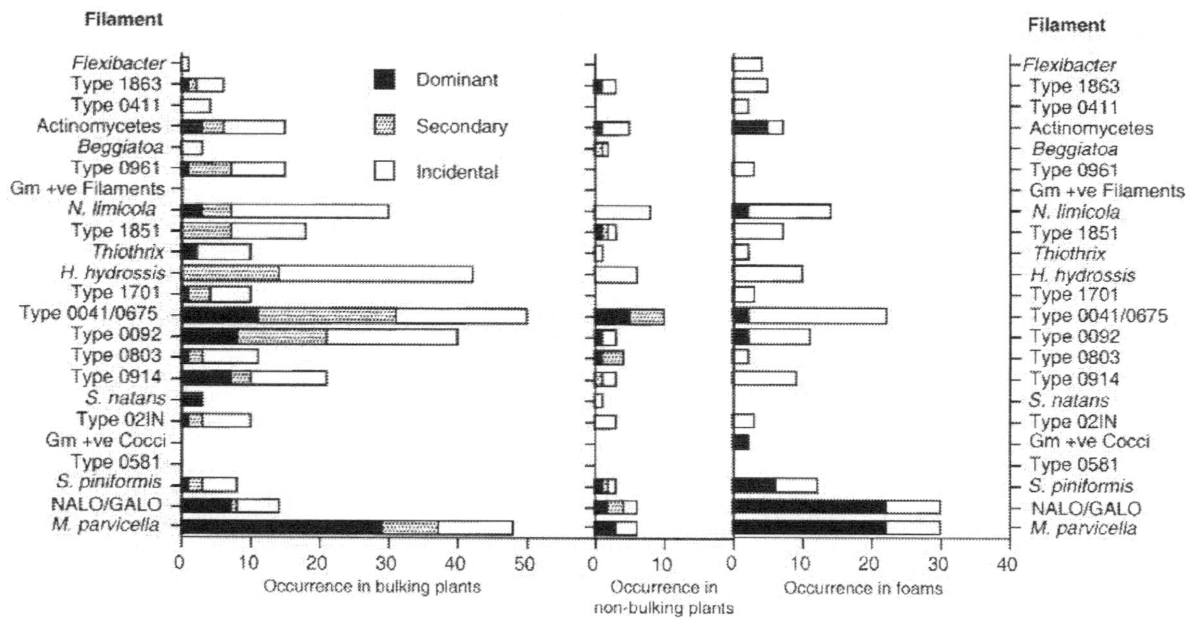

Fig. 6.2: Survey data of Bradford (1998) of bulking and foaming plants in Queensland, showing major filamentous bacteria detected.

6.2.4 Do the filamentous populations in activated sludge plants change with time?

The problems inherent in drawing conclusions based on a single sample have been mentioned already. It would be expected that the filaments in plants would change with changing conditions, and the available data partially support this. For example, Knight *et al* (1994) followed the filaments present during start up of a modified UCT plant in Bendigo, Victoria in the first 10 months of its operation. Some filaments were always seen (eg Type 0041/0675) in the mixed liquor while some appeared sporadically (eg *Halisconemobacter hydrossis*). Gradually *M. parvicella,* Type 0092 and Type 0914 became dominant, and 7 years after the plant started these filaments are still the abundant members of the community for most samples examined, so fairly stable communities may persist.

In the foam, which is ever present, GALO (probably mainly *G. amarae*) were always seen in large numbers, with *M. parvicella* and Type 0092 gradually appearing as they became dominant in the mixed liquor, and Type 1851 also present. It was not clear why these population changes occurred, but possible reasons will be discussed briefly later.

6.2.5 Which operational factor affect filamentous populations and consequently determine incidents of bulking and foaming

It should be clear from the survey data presented here, that most of this work is descriptive, and not especially useful in helping us determine which conditions favour the proliferation of certain filaments, and induce or otherwise these operational problems. There are several reasons for this, most of which are not peculiar to Australia. The main difficulty is the number of uncontrolled variables associated with activated sludge processes which makes it almost impossible to identify and isolate a single causative factor. The other problem is that the data collected, even in plants where relevant plant parameters are regularly monitored (eg SVI, DSVI, sludge age, etc) is often insufficient to draw any sound conclusions. This problem will vary from plant to plant depending on the skills of the operator and the resources available to do such analyses (and in Australia these are decreasing rapidly as cost cutting measures), but very few, except the largest plants carry out these tests frequently enough to make the data useful. Furthermore only a very small number in Australia carry out any microscopic examination of mixed liquor and foams to identify the filaments present.

Most of the available data relate to foaming, although Seviour *et al* (1994) did suggest that many of the rules of thumb described by both Eikelboom and van Buijsen (1983) and Jenkins *et al* (1993) for controlling bulking filaments did not always appear to apply to most plants, (conventional and EBNR) in Australia. For example, many of the common bulking filaments like *M. parvicella*, Type 0041/0675, Type 0914, Type 0803 and *Nostocoida limicola II* were seen in both under and overloaded plants operating at both long and short sludge ages and over a wide range of operating temperatures. They suggested that such data might reflect widely different physiological attributes in a single morphotype (something not detectable with oligonucleotide probes), and stressed the need for caution in making any generalisations about the factors which seemed to favour the proliferation of certain filament morphotypes, which provide the present bases for their control (Eikelboom & van Buijsen 1983; Jenkins *et al* 1993; Wanner 1994 a,b). These problems, it seems to us will only be overcome by studies with pure cultures of these bacteria; work currently proceeding in our laboratory. However Seviour *et al* (1994) did show Type 1863 was more common in short sludge age plants, and suggested that both bulking and foaming incidents seemed more common in under loaded plants. This generalisation was supported by Bradford (1998) in her survey of Queensland plants, although Blackall *et al* (1991) also with Queensland plants, proposed the opposite. These latter two studies also disagreed on the occurrences of foaming in conventional as compared to oxidation ditch systems, where Blackall *et al* (1991), recorded the former with a higher incidence of foam, and Bradford (1998) the latter. Such results (and our unpublished surveys confirm this) only serve to illustrate the difficulty in reaching generalised conclusions on cause and effect relationship in systems with so many variables, and usually with inadequate data available, and in our view render any such views almost meaningless.

6.2.6 Control Measures for Bulking and Foaming

Clearly unless the complex problems mentioned in the preceding discussion are overcome, i.e. sorting out the biodiversity among the filaments, understanding the factors which favor or otherwise their growth and setting up controlled systems where their influence can be determined, all attempts to control both bulking and foaming are empirical and with no guarantee of being successful. It is not possible to translate the experiences gained in one plant or one country to another, unless it is certain that the bacterial populations causing the problem are identical in both. The application of molecular probes will assist in this long term objective. Also in our experience, biomass always contains a mixture of several different filamentous bacteria, and so it may be possible to reduce the contribution of one type, but in so doing, some of the other filaments may then in turn become dominant and cause equally troublesome problems. Unfortunately few plants in Australia monitor their filament populations. So like other aspects of this review, our reservations on the value and meaning of the survey data are strong, and need to be interpreted with caution.

What can be said is that the studies which have been done suggest, not too surprisingly in view of the above comments, that there is no universal control measure for either bulking or foaming. In fact Blackall *et al* (1991) demonstrated that none of those used in the Queensland plants they surveyed were effective for foam control. These measures included many suggested by overseas experience to work, like changing the RAS recycle rate, altering (ie decreasing) sludge age, varying the levels of MLSS and addition of a commercial preparation claimed to remove foam. Some of these e.g. reducing sludge age are based on washing the foaming bacteria, which incorrectly have been assumed to always be slow growing bacteria (Soddell 1999), from the system. Selectors (Jenkins *et al* 1993; Wanner 1994) are installed in some plants, but not many, and how effective they are is not known to the authors.

We know of no survey undertaken in Australia to determine the methods used to control bulking, but much of what was said about foaming also applies to bulking. Thus increasing the wastage rate of sludge or chlorination are the common knee-jerk reactions to an episode of bulking (which is often defined differently by different plant staff), and rarely if ever, are the filamentous populations checked microscopically before, during or after.

The case history described next will illustrate these approaches and problems.

6.2.7 An Australian Case Study

The plant discussed here is a conventional activated sludge system which caters for a drainage area with a total population in excess of 200,000 people. It treats mainly domestic wastewater, with a small proportion of light industrial waste. This system consists of three interconnected plants, the first (Plant A) was built in 1932 (6102 m^3), the second (Plant B) in 1956 (5728 m^3) and the third (Plant C) became operational in 1976 (5680 m^3). All three treat the same raw influent waste roughly in proportion to their relative sizes at a rate of 50 - 75,000 e.p.

6.2.7.1 Plant Design Features

All three receive waste which has passed through primary clarification, although the clarifier in Plant A is considered to be more efficient than those in plants B and C in removing fats and grease.

Plant A has a buffled serpentine (plug flow) aeration basin, and a 6.1 Ml day $^{-1}$ flow rate. The secondary clarifier has a double scraper system, and a 40° slope on the tank floor. This plant only very occasionally encounters bulking or foaming, and the average DSVI is 150 mL g^{-1} (the range during the monitoring period was 96 mL g^{-1} – 245 mL g^{-1}).

Influent flow to Plants B and C is typically 5.7 Ml day^{-1}, and both have a slightly different configuration to Plant A, operating more as completely mixed systems. Both have a secondary clarifier with a single scraper, and a floor angle of between 20° and 30°. Both suffer from episodic bulking and foaming. The average SVI and DSVI ranges over the sampling period were 123 and 71–353 mL g^{-1}, 141 and 57-396 mL g^{-1} for plants B and C respectively. Both SVI and DSVI data are recorded by the operations staff. However SVI data are more complete, and have been used to determine the biomass settling characteristics listed below.

The D.O. probes are situated at the head of Plant A but at the ends of Plants B and C until recently, when they were also moved to the heads of these plants. Recorded DO data reveal that these set levels were maintained at or near the set points at these monitoring plants. However, it is clear from examining the effluent ammonia - N and nitrate - N data that substantial variations in the aeration efficiencies or aerated mass fraction occur in all three plants. None completely nitrifies, and N - mass balance data indicate that a high rate of denitrification also takes place, albeit sporadically, in all three. Plant A typically nitrifies the least efficiently, (less than 50% conversion), while Plants B and C typically show c. 80% and 60% conversion of ammonia to nitrate.

6.2.7.2 Microscopic Features of the Biomass

This work was carried out in our laboratories on weekly samples viewed within 24h of their collection, as no regular microscopic examination is performed by the operations staff at this plant. The major filament Types (Eikelboom and van Buijsen 1983; Jenkins *et al* 1993: Seviour & Blackall 1999) were identified and filament abundances subjectively assessed.

All three plants had a distinctive biomass, in terms of floc structure and size, and their main filament Types. These changed over the 6 month sampling period, but usually only ephemerally.

Plant A had characteristically small compact flocs with good filament backbone, but little or no interfloc bridging. The dominant filament seen was *Nostocoida limicola II*, which has been cultured and its 16S rDNA sequence determined. Towards the end of the monitoring period, this organism became almost the only filaments seen in this biomass and biomass viability also appeared to decline. The only other filaments seen regularly were Type 021N and Type 0041/0675, and occasionally, *Haliscomenobacter hydrossis*.

In Plant B, the biomass changed dramatically in the 6 months. Filament numbers increased from a few per floc to more than 20. The presence of protozoa was uneven, and often none could be seen over a several week period. Floc size and shape was much more variable than in Plant A, although the flocs were larger. Type 021N was the dominant filament during this period, although *N. limicola II* and Type 0041/0675 were also present as secondary filaments.

The biomass in Plant C also contained large flocs and extensive interfloc bridging was common, so bulking in this plant was frequently encountered. The troublesome filament was Type 021N, reported in many countries as a bulking filament. However in both Plants B and C, when foaming occurred, the major filament present in the foam was also Type 02IN. In Plant C, as in Plant B, Type 0041/0675 and *N. limicola II* were always seen as secondary filament Types. *Haliscomenobacter hydrossis* was also present.

It is not know whether Type 02IN is the primary foaming organism or one which is transported into foams by some other mechanism/s. It is not strongly hydrophobic, according to the MATH assay, suggesting it may not be a causative foam foaming filament.

6.2.7.3 Bulking and Foaming Incidents

DSVI and SVI data are used in this plant to determine if bulking is occurring; bulking is considered to occur when the DSVI exceeds 150 mL g^{-1}. Control is by chlorination, which incidentally is not the universal method of choice for all plants in Australia. In many chlorination is seen as highly undesirable, and only used as a last resort. Unfortunately, during the study period, no record of when chlorine was added was kept. The target DSVI used is 150 mL g^{-1}, which after chlorination usually takes 2 - 3 days to reach, and often in Plant C, there is an overshoot where the DSVI drops as low as 80 mL g^{-1}. Chlorine is injected at a constant rate of 70 kg day^{-1} for Plants A and B, and 60 kg day^{-1} for Plant C into the RAS line of each basin according to the recommendations of Jenkins *et al* (1993). The plant layout is such that only one basin at a time can be chlorinated. Plant C is chlorinated with care at the lower dose level, as the biomass otherwise appears to fragment to a loose, poorly structured floc. Any remaining foam also tends to disappear after chlorination.

The DSVI data show that bulking was infrequent in Plants A and B, while Plant C showed a cyclic pattern of bulking with a frequency of about 4 - 6 weeks.

No microscopic/microbiological monitoring of the biomass is carried out after chlorination to test its efficiency, and DSVI values are the only monitoring used in this control strategy. In our experience, this is the usual situation in Australian plants, and some plants carry out no subsequent monitoring after chlorination or other control methods.

6.2.8 Conclusions

1) Several surveys have been carried out into describing the major filamentous bacteria in both conventional and EBNR activated sludge plants.
2) Australian plants contain most of the filament morphotypes seen in plants in other parts of the world. *Skermania piniformis* is a common foaming bacterium whose presence has not been reported commonly elsewhere, even though light micrographs clearly show it occurs in USA, and in Czech Republic (Wanner, personal communication), and probably elsewhere.
3) There is some evidence that *M. parvicella* is seen less often in plants in warmer states e.g. Queensland. However, it is the major bulking and foaming filament in Australia.
4) It is virtually impossible in our view to identify 'cause and effect' relationships from the studies carried out in Australia, as too many interacting and poorly understood variables are involved.
5) Control strategies for bulking and foaming almost without exception rely on no prior identification of the filamentous bacteria present in the biomass or foam. Foaming is easily recognised, and either SVI or DSVI values are used to assess whether bulking is occurring.
6) We know of no survey data for bulking control in Australian plants. In some chlorination is used, and in others reducing sludge age, changing aeration or selectors have been tried. In almost all cases, these are carried out empirically with no prior filament identification, and often monitoring the effect of the selected control strategy is haphazard, and again does not involve any microscopic examination of the biomass.

6.3 CZECH REPUBLIK (I. RUZICKOVA, O. KRHUTKOVA, J. WANNER)

6.3.1 Separation problems - situation up to 1980s

Although some of the first papers on identification of activated sludge filamentous microorganisms came from (former) Czechoslovakia (Cyrus and Sladka, 1970; Sladka and Ottova, 1973), a systematic monitoring was not performed in this country at that time. There were several reasons for this:

- activated sludge plants built and operated up to the mid-1980s were either conventional, rather high loaded plants (Θx = 4-6 d) or extended aeration plants with aerobic sludge stabilisation ($\Theta x > 25$ d); in both extremes the activated sludge plants did not exhibit significant sludge separation problems which could attract the attention of either plant owners or researchers
- no team in the country was equipped, supported and skilled enough
- serious sludge separation problems started at the end of 1980s and beginning of 1990s when new biological nutrient removal activated sludge plants were introduced in a short period of time.

Although no official national survey existed, it can be concluded from examinations of numerous activated sludge that the commonly occurring Types were *Sphaerotilus natans*, Type 021N, *Thiothrix* spp., Type 0041/0675, *Haliscomenobacter hydrossis*, *Nostocoida limicola*, *M. parvicella*, nocardioforms and fungi. Severe bulking problems were not frequently reported in municipal plants and the bulking problems usually were connected with the seasonal food industries in many towns (e.g. fruit and vegetable canneries, sugar beet mills, wine grapes processing, distilleries). Wastewaters from breweries, (in all major Czech towns), caused permanent presence of filamentous microorganisms in activated sludges.

Foaming in activated sludge plants treating domestic wastewaters was virtually unknown until the late 1980s. The first serious case was in an anaerobic - oxic system (EBPR); the microorganisms found in the foam were *Nostocoida limicola* and Type 0041 (Wanner and Grau, 1989).

6.3.2 Separation problems - situation up to mid-1990s

In the 1990s, a shift in filamentous microorganisms dominance was observed at many plants. The Prague central plant is a well-documented example because its activated sludge has been used for more than 15 years as an inoculum for experimental units.

The plant has a plug-flow system aeration basin and filamentous microorganisms were always present. In the 1980s *Sphaerotilus natans* and/or Type 021N, were dominant (Wanner et al., 1987; Wanner et al., 1988) however, in 1991 and 1992 *Nostocoida limicola* had become dominant and *Sphaerotilus natans* and Type 021N levels were usually negligible. Because no changes had occurred the shift could be attributed to changes in the wastewater.

At this time, foaming was observed in conventional activated sludge systems treating domestic sewages. The causative microorganisms were Nocardioforms. Perhaps the biological foaming was related to new detergents introduced into the Czechoslovak market after 1990 and to a shift in the household consumption of fats from animal to cheaper vegetable.

6.3.3 WWTPs screening - 1995 - 2000

Regular activated sludge examination started in the early 1990s at plants in the Prague area and then spread survey was then extended to the whole country. The aim of this study was to determine the most common filamentous microorganisms in Czech plants and to study their effects on activated sludge flocs and on their separation properties.

6.3.3.1 Methodology

In early 1995 a questionnaire was mailed to supervisors at most activated sludge plants in the country, to collect the answers to the following questions:

- design and current flow of plant; sludge age; mixed liquor concentration; SVI and average summer and winter temperature
- process type and schematic configuration
- wastewater type (sewage, municipal wastewater, type of industrial wastewaters), average composition (COD, BOD, SS, N and P) and required effluent quality
- for upgraded WWTPs data on changes in process facilitie, settling properties, etc.
- separation problems: if any, SVI data, behaviour of sludge in secondary clarifier, microscopic picture of the sludge
- for sludge bulking: extent of problem, counter-measures taken (selection from a list)
- for foaming: description of observed problems, counter-measures taken (selection from a list)
- seasonal changes in separation problems.

From questionnaire responses, plants with persistent separation problems were selected, then visited to verify the data from the questionnaire response and to collect mixed liquor and foam samples for microscopic examination. Activated sludge samples were taken at the effluent end of the aeration basin, while foam samples were taken from the surface of the aeration basin.

6.3.3.2 Microscopic analysis and sedimentation properties evaluation

Microscopic analyses of samples were done by the method of Eikelboom and van Buijsen (1981) and Jenkins *et al.* (1993) for:

- floc characteristics and effect of filaments on floc structure
- protozoa and *Zoogloea*
- overall and individual filamentous microorganisms abundance
- photographic documentation and videorecording of characteristic features of observed filaments for training and educational purposes.

Filamentous microorganism abundance was rated according the scale 0-6; the filament occurrence was rated dominant when the individual abundance was 4 (more than 5 filaments per floc - very common) or higher.

Sludge sedimentation properties were measured by SVI. Activated sludge was classified by SVI value using the following: "well settling" with SVI up to 100 mL/g, "light", with SVI 100-200 mL/g and "bulking" with SVIs higher than 200 mL/g.

6.3.4 Results of separation problems and filamentous micro-organisms screening

In the years 1996 and 1997, 78 and 72 activated sludge plants respectively were visited and sampled, respectively (Wanner *et al.*, 1998 and Wanner *et al.*, 2000).

Bulking was not such a critical problem as in the past perhaps because most contemporary plants were designed using bulking control recommendations (Wanner, 1994a).

These designs have been efficient in suppressing filaments from Groups S and C (Wanner and Grau, 1989) and in this survey it was difficult to find filaments like *Sphaerotilus natans* or Type 021N, in municipal WWTPs. The survey demonstrated that in the past 10 years filaments from Group A (*M. parvicella* and *Nostocoida limicola*) have become dominant. The elimination of Group S and C filamentous microorganisms from BNR systems can be explained by metabolic selection to which Group A filaments is less susceptible.

6.3.4.1 Filamentous micro-organism abundance

The abundance of dominant filaments exhibits an increasing tendency as illustrated by Figure 6.3: in 1996 abundance 4 and 5 were almost equal while in 1997 abundance 5 clearly dominated (in smaller extent similar to abundance 6). The quality of activated sludge flocs affected the settling properties at medium and low abundance of filaments whereas high filament abundance always resulted in bulking or foaming regardless the flocs characteristics.

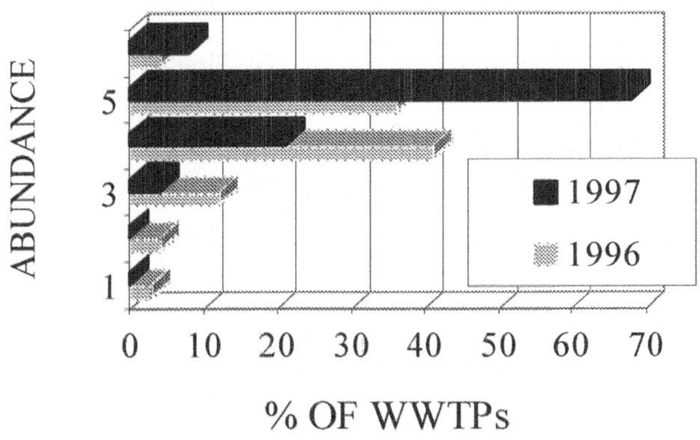

Fig. 6.3. The difference in abundance of dominant filaments at 78 (1996)/72(1997) activated sludge plants in the Czech Republic between 1996 and 1997.

6.3.4.2 Dominant filamentous microorganisms in sludge and foam

The commonest dominant filament in conventional and BNR plants were *M. parvicella*, *Nostocoida limicola* and Types 0092 and 0041/0675.

The national survey showed that foaming by filamentous microorganisms (Group F) (Wanner and Grau, 1989) is becoming the most important solid separation problem. Table 6.1 shows the most common foam-forming filaments in plants with persistent foaming. Foams were dominated by *M. parvicella*, actinomycetes GALOs (previously known as NALOs – *Nocardia amarae* - like organisms) and by *Nostocoida limicola*. Table 6.2 shows that foaming was common in all activated sludge modifications and being most pronounced in plants with C and N removal (denitrification - nitrification). Fewer foam problems existed in conventional plants.

The survey determined foam control strategies (Table 6.3)

Table 6.1. Most common filamentous microorganisms in biological foams (1996 and 1997)

1996			1997		
Filamentous microorganism	Number of plants	%	Filamentous microorganism	Number of plants	%
M. parvicella	16	73	*M. parvicella*	18	55
GALOs	6	27	GALOs	12	36
N. limicola	5	23	*N. limicola*	4	12
No. of foaming plants	22		No. of foaming plants	33	

Table 6.2. Correlation of activated sludge process type and frequency of biological foaming frequency (1997), C - organic carbon removal; N - nitrogen removal; P - EBPR

Type of the process	Number of plants	Foam occurrence	%
C	51	19	37
C, N	13	9	69
C, N, P	8	5	63

Table 6.3. Most common foam control strategies in the Czech Republic (1996 - 1997)

Foam control strategy	Number of plants	%
Water sprays	26	42
Skimming /mechanical foam removal	18	20
Return activated sludge manipulation	16	26
Sludge age reduction	15	24
Installation of baffles	13	21
Initial contact zones ("selectors")	10	16
Chlorination of foam	3	5

6.3.5 Development of filamentous population in Czech activated sludge plants between 1997 - 1998

Of the more than 70 activated sludge plants sampled in year 1997, 33 WWTPs were selected for visitation in 1998.

6.3.5.1 SVI and total abundance of filaments

Table 6.4 shows that most sludges examined were "light" in both 1997 and 1998. Bulking sludges increased from 21 % in 1997 to 36 % of bulking sludges in 1998. High filament abundances did not necessarily mean high SVI (Wanner, 1994b). In some cases it was found that firm and compact floc morphology and an excessive quantity of filaments resulted in good settling while in others poor floc structure with low filament levels led to poor settling.

Table 6.4. Evaluation of total filament abundance and settling properties in 1997 and 1998 (33 WWTPs)

Year	SVI (ml/g)	Total abundance of filaments				
		3	4	5	6	Σ**%
1997	< 100	1	4	6	1	36
	100 - 200	1	2	11	-	43
	> 200	-	-	7	-	21
	Σ*%	6	18	73	3	100
1998	< 100	-	1	2	1	12
	100 - 200	1	3	11	2	52
	> 200	-	-	11	1	36
	Σ*	3	12	73	12	100

* total % of activated sludge samples with overall filament abundance

** % WWTPs in indicated SVI range

During the two screening years the mixed liquor samples could mostly be described as follows: small-medium flocs with a few compact centres, firm texture and irregular shape, filaments frequently present, affecting floc structure by bridging and by causing open floc structure.

6.3.5.2 Dominant filamentous microorganisms in activated sludge mixed liquor and foams

This survey covered both conventional and BNR plants, i.e. plants, incorporating anaerobic and/or anoxic zones. 18 (55 %) were conventional COD removal plants only, 8 WWTPs (24 %)

had a denitrification zone and 7 plants (21 %) were BNR systems. In both years the four most frequent dominant filaments in mixed liquors were *M. parvicella*, *Nostocoida limicola*, Type 0092 and Type 0041/0675, and in 1998 *Haliscomenobacter hydrossis* was also seen (Figure 6.4). The most frequently observed filaments were the same observed in all WWTP configurations.

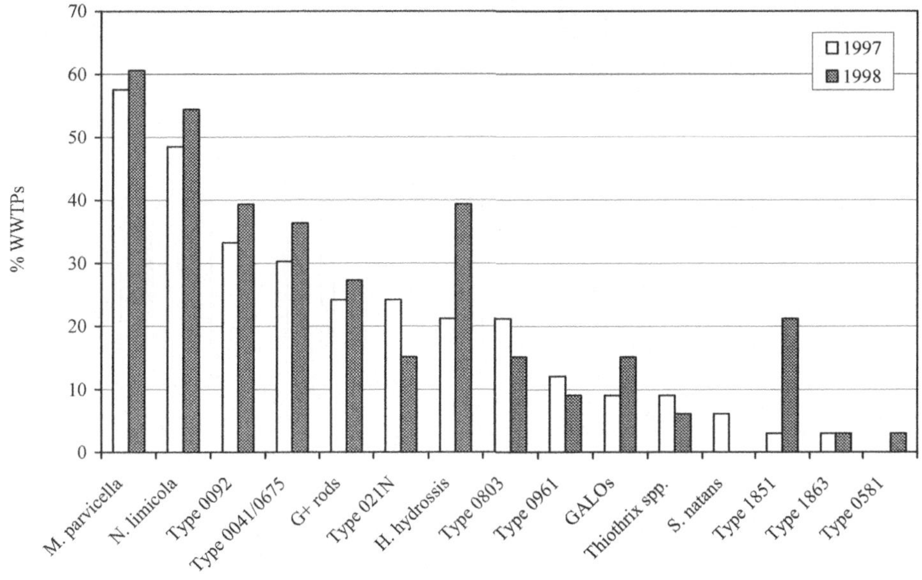

Fig. 6.4. Comparison of distribution of the dominant bacteria in activated sludges in 1997 and 1998

Figure 6.5 shows the population distribution of secondary filaments. *Haliscomenobacter hydrossis*, Type 0041/0675 and Type 0803 were the commonest secondary microorganisms in 1997. In 1988 *Haliscomenobacter hydrossis* and Type 0041/0675 were the commonest secondary filaments. In addition, Types 0961, 1701 and *Thiothrix* spp. were common secondary filaments in 1998.

In 1997 and 1998 13 biological foams (39 % of all WWTPs) and 26 biological foams (79 % of all WWTPs) were observed at the selected WWTPs. Thus foaming problems were more pronounced in the year 1998. Foams in aeration basin corners were mostly commonly observed. Figure 6.6 shows that, *M. parvicella*, Gram + rods (not evaluated in 1997) and GALOs were the most frequent foam-formers in both years. An increase of biological foam quantity containing Types 1863, 1851 and 0581 occurred. Other filaments were found in foams but they did not appear to support foam production because of their low abundance. Six types of filamentous bacteria were numerous in foam: *M. parvicella*, GALOs, *Nostocoida limicola*, Types 0092, 0041/0675 and 1851.

Unidentified Gram + rods were common dominant microorganisms in mixed liquor and foam in both years. The rods were Gram +, Neisser-negative and most contained Neisser-positive granules. They grew as individual cells or very short chains (2 - 4) of cells of about 0.8 μm in diameter. It is not known if these bacteria are a filamentous microorganisms growth form (*M. parvicella*, GALOs) or a undescribed bacterial species. The G + rods were mostly found with both of the above mentioned filaments.

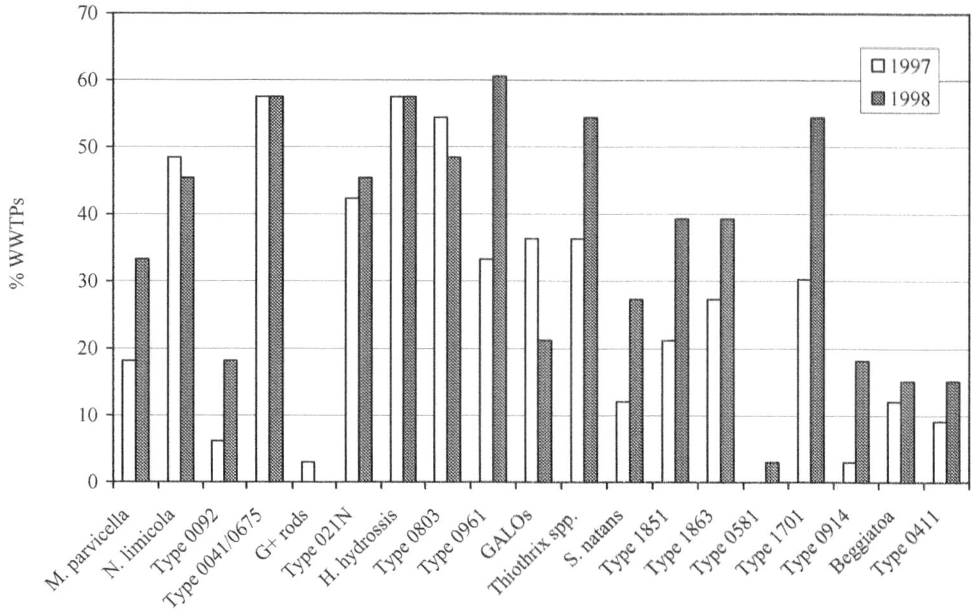

Fig. 6.5. Comparison of distribution of the secondary bacteria in activated sludge in 1997 and 1998

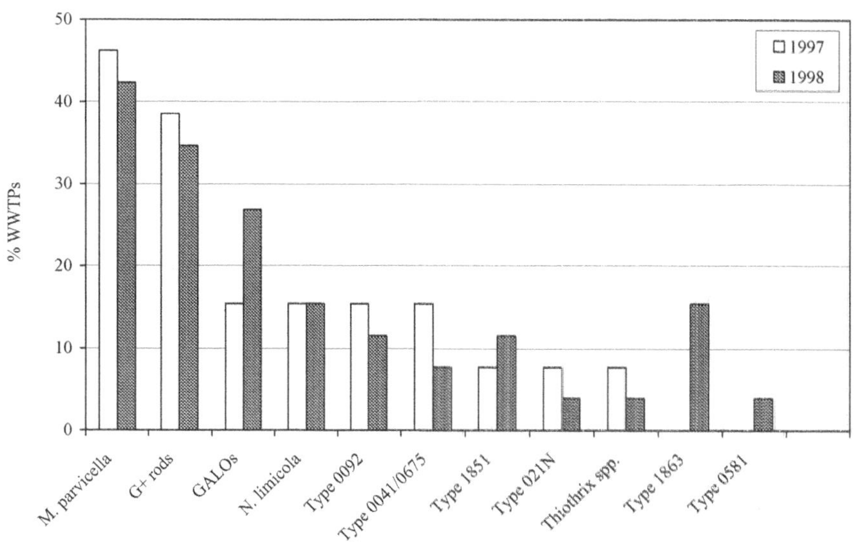

Fig. 6.6. Comparison of distribution of filaments appeared in foams in 1997 and 1998 (some samples of foam were created by more than one dominant filamentous type)

6.3.6 Screening of eight nutrient removal plants in 2000

6.3.6.1 Characterization of monitored WWTPs

All monitored WWTPs were BNR plants operated at low organic loads. Four were R-AN-D-N systems (regeneration - anaerobic - anoxic and oxic zones; Kos *et al*., 1992) with presettling and corridor-type basins, other four were AN-D-N systems. The wastewaters were mostly of domestic sewage and wastewaters from food (milk, brewery, meat, preserving and starch industry) or tanning industries.

Each of the eight plants was microscopically examined at least four times a year (at least one sample of mixed liquor and one sample of foam per season).

6.3.6.2 Activated sludges and settling properties

Flocs in all 32 sludge samples had firm texture and zoogloeal colonies did not affect their settling properties. Higher micro-organisms were equally distibuted, among flagellates, amoeboid protozoa and stalked ciliates. No apparent relationship existed between floc characteristics, plant type and season (Krhutkova *et al.* 2001).

Most sludge samples were classified as "light" (SVI: 100-200 mL/g). Floc structure, filamentous type and location, quantity and density of poly-P (PAOs) clusters, and presence of inorganic salts had a significant influence on activated sludge settleability. There was no clear correlation between SVI values, filamentous abundances and actual activated sludge settling properties.

The total abundance of filamentous bacteria was never < 3 in the mixed liquor and was most frequently 5 (more than 20 filaments per floc).

6.3.6.3 Dominant filamentous microorganisms in activated sludge mixed liquor

Survey results from 2000 were: Type 0092 and *M. parvicella* were the most common dominant filaments, *Haliscomenobacter hydrossis* and Type 0041/0675 were secondary filaments always found at low abundance (1 - 3) at all operating conditions.

At some plants P removal by chemical addition (Fe salts) was practised. Three plants used ferric sulphate addition. The influence of chemical precipitation on filament types on settling properties is shown in Table 6.5.

At WWTPs without chemical precipitation there were 6 samples of bulking sludge containing: in descending order *M. parvicella* (67 %), Type 0092, Type 0803 and *Nostocoida limicola* (33 %). Type 0092 was identified in all 10 samples of well settling or light sludge. *M. parvicella* was present in 30 % of the samples. Type 0092 occurred in 16 samples where inorganic salts were added; its appearance was comparable to the samples without iron salt addition, about 80 % in good settling sludges and 33 % in bulking samples. *M. parvicella* population was mostly found in "well" settling sludges (80 %). In samples with higher SVI values it was present 65 % of the time. Thus the overall dominant presence of both Type 0092 and *M. parvicella* was in plants with and without chemical precipitation. The influence of *M. parvicella* on settling may have been partially eliminated by the addition of iron salts.

Table 6. 5. Correlation between SVI values, filamentous Types and chemical precipitation

	Addition of $Fe_2(SO_4)_3$			No Addition of $Fe_2(SO_4)_3$		
SVI (ml/g)	< 100	100 - 200	> 200	< 100	100 - 200	> 200
Type 0092	80 %	75 %	33 %	100 %	100 %	50 %
M. parvicella	80 %	62 %	67 %	25 %	33 %	67 %
Type 1851	-	-	-	-	-	33 %
N. limicola	-	-	33 %	-	-	33 %

6.3.6.4 *Biological foams*

In 2000 there was no biological foam at 3 WWTPs and in 2 of them biological foams were always present. Seasonal shifts in foaming occurred. Most foams appeared in winter and spring, while biological foams were observed at 2 plants only after warm weather. Table 6.6 shows that, the foam-forming filaments were mainly *M. parvicella* and Gram + rods, and some Types 0092, 0581 and GALOs. Well settling activated sludge formed foam less frequently than light bulking sludges.

Table 6.6. Occurrence of foam-forming filamentous types at eight WWTPs during year 2000

	(R-AN-D-N)				(AN-D-N)			
	1	2	3	4	5	6	7	8
Winter	-	Type 0581	*M. parvicella* Type 0092	-	*M. parvicella*	-	*M. parvicella*	Gram + rods
Spring	-	Type 0581 Gram + rods	*	-	*M. parvicella* Gram + rods	-	*M. parvicella*	*M. parvicella* Gram + rods
Summer	-	-	*M. parvicella* Gram + rods	-	*M. parvicella* Gram + rods	-	GALOs	*M. parvicella*
Autumn	-	-	*	-	*M. parvicella*	-	GALOs	-

* unstable and sleazy foam was observed, sample was not taken

6.3.7 FOAM CONTROL STRATEGIES

6.3.7.1 *Water sprays*

Biological foams can be destabilized by dilution. The amount of water sprayed on the foam needs to be much higher than in water sprays used for detergent foam control. Spraying with water does not eliminate the cause of foaming. The filaments from the collapsed foams are returned to the mixed liquor or RAS so that this method retains the filaments in the system. Thus, the use of water sprays should only be considered an immediate emergency measure especially when a massive amount of floating biomass escapes the secondary settling tanks. The installation of water sprays in mixed liquor channels (Figure 6.7) or on final clarifier bridges was an efficient method for reducing biomass losses at several WWTPs.

Fig. 6.7. Water sprays in mixed liquor channels

6.3.7.2 Skimming (mechanical removal) of foam

The mechanical removal of foam prevents foam-forming microorganisms from taking advantage of the prolonged biomass retention time in the foam. The latest secondary clarifier design recommendations is to equip both rectangular and circular settling units with skimmers which act over the whole water surface (Figure 6.8) and with foam traps (Figure 6.9). Biological foam can be trapped and mechanically removed in mixed liquor channels (Figure 6.10). The foam removed from these traps or by the skimmers should be handled separately from WAS, especially when the WAS is stabilized in anaerobic digesters.

Fig. 6.8 Skimming of secondary settling tank surface

Fig. 6.9. Foam trap with final clarifier tank surface rotating bridge

Fig. 6.10. Example of biological foam mechanical removal

6.3.7.3 Return activated sludge manipulation

Increased RAS flow rate can reduce retention time in the thickened bottom sludge blanket preventing denitrification that helps float sludge. This countermeasure should not be used on a long-term basis.

6.3.7.4 Manipulation of SRT

The SRT required for reliable elimination of foam-forming microorganisms (especially GALOs) from activated sludges are low and may negatively affect the treatment efficiency. The low SRT consortium is dominated by fast-growing organotrophs, so that BNR is not possible.

6.3.7.5 Additional installation of baffles

Proper secondary clarifier design and construction is essential for foam control. Secondary clarifiers should be able to retain and collect floating biomass and prevent its escape into the effluent. Final clarifiers should be equipped with baffles that protect the effluent weirs and with efficient skimming devices: without these a low SS effluent cannot be produced (Figure 6.11 and 6.12).

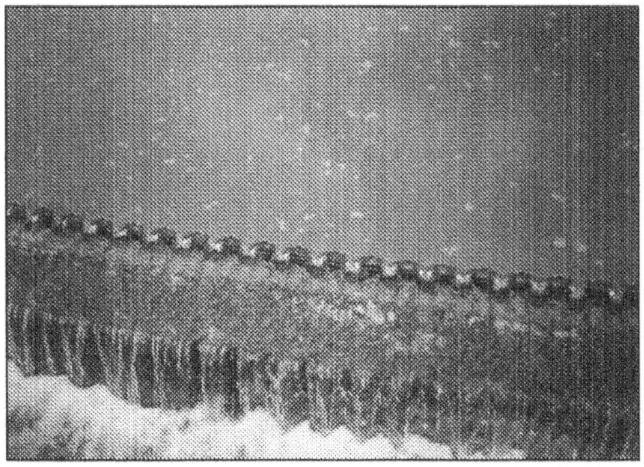

Fig. 6.11. Small patches of floating biomass

Fig. 6.12. Massive escape of biomass

6.3.7.6 Initial contact zones ("selectors")

Methods such as kinetic or metabolic selection are less successful for foaming control than for bulking control (Table 6.3) because the selection pressure that can be achieved is not sufficient to suppress the growth of filaments like *M. parvicella*. Often readily biodegradable organic substrate penetrates through anaerobic and/or anoxic zones into the oxic zone where it is used by filaments at a higher rate than in non-aerated zones. Extensive foaming problems were not found in activated sludge plants with well-established EBPR.

6.3.7.7 Chlorination of foam

Control of foam-forming filamentous microorganisms by chlorine is not widely used. When chlorination is used in a BNR plant a temporary deterioration of nitrification and EBPR efficiency is to be expected.

6.4 DENMARK, GREECE AND NETHERLANDS (K. ANDREASEN, A. ANDREADAKIS, D. EIKELBOOM)

6.4.1 General situation

In Denmark, Greece and The Netherlands an intensive upgrading of municipal wastewater treatment plants to BNR has been carried out during the last decade. Stringent effluent standards have required more effective sludge separation in the final clarifier, so that well settling activated sludge is a key parameter. In many BNR activated sludge plants sludge settleability deteriorated after the BNR upgrade, often due to increased filamentous microorganisms levels (Blackbeard et al, 1988; Eikelboom 1993; Kristensen et al., 1994; Andreasen and Sigvardsen 1996).

Because of this, the influence of filamentous microorganisms on the sludge settling properties in BNR activated sludge plants was investigated in an EU project conducted between 1993 and 1996. More than 200 BNR plants in the three countries were surveyed and information on DSVI, filament index, dominant filamentous organisms and plant layout were obtained. From these data, it was possible to assess the magnitude of the bulking problem in BNR plants, and identify some of the controlling factors. The EU project also included laboratory, pilot and full scale studies, that provided additional information on which to base control strategies. In this chapter we present the major findings of the EU project together with additional experiences from these three countries.

6.4.2 Wastewater composition

Information connected on wastewater composition and findings from other investigations (e.g. Henze et al., 1995; Andreadakis et al., 1998) give the general picture shown in Table 6.7. There are not any major differences in wastewater composition in the 3 countries; only wastewater temperatures are different. Infiltration of ground water and industrial discharges produce significant variations in composition from the values presented in Table 6.7. Only limited information is available on the specific organic compounds that make up the COD and BOD values presented in table 6.7. Raunkjær et al. (1994) found that in 4 Danish wastewater treatment plants and proteins, carbohydrates, lipids and VFAs made up respectively 28%, 18%, 31% and 6% of the total COD. Studies in Greece (Andreadakis et. al.,1998) showed that about 30% of the

COD was soluble and its readily biodegradable fraction ranged between 6-14% (corresponding to a VFA fraction on the order of 4-10%). Both of these fractions depended on the residence time of the wastewater in the sewerage system. In Dutch wastewater treatment plants about 1/3 of the influent COD was soluble.

In Denmark, only approximately 50% of the organic load on wastewater treatment plants is domestic; the majority of the remaining load comes from food processing industries. In both The Netherlands and Greece industrial wastewater contribution to the municipal wastewater treatment plants is generally lower. A significant portion of the domestic load in Greece (5-20%) is from septage.

Table 6.7. General characteristics of the wastewater in the three countries

	Denmark*	Greece**	The Netherlands***
COD, mg/L	530	500	550
BOD, mg/L	250	250	200
SS, mg/L	300	300	?
TN, mg/L	50	50	49
TP, mg/L	16	15	8
pH	7-8	7-8	7-8
Temp. °C	8-18	15-28	5-20
Protein % COD	28		
Carbohydrates % COD	18		
Lipids % COD	31		18-41
VFA % COD	6	4-10	5-10

(*Henze et al., 1995; ** Andreadakis et al., 1998; ***Raunkjær et al., 1994)

6.4.3 Process configurations

Most of the wastewater treatment plants in the three countries are based on single-stage activated sludge systems, in which the COD, N and P removal is performed in the same stage. Primary sedimentation is often included in the larger plants (> 50.000 PE). An overview of the process configurations can be found in Eikelboom et al. (1998).

Nitrogen removal is normally accomplished by predenitrification or by simultaneous/intermittent denitrification, without the addition of an external carbon source.

The configurations include:

- Various configurations in which nitrate from oxic basin is recirculated back to a denitrification basin.
- Alternating methods (e.g. BioDenitro), where the flow patterns and the function of the process tanks alternate
- Simultaneous nitrification and denitrification in the same basin (e.g. Carrousel systems).
- Intermittent denitrification, where denitrification is achieved by intermittent aeration.

For the plants with intermittent and simultaneous denitrification part of the readily biodegradable substrate may be fed directly to the aerobic zone, while all of biodegradable substrate is normally utilised in the anoxic zone in recirculation configurations and BioDenitro plants.

P removal is achieved either by addition of chemicals (iron sulphate, iron chloride, aluminium sulphate or polyaluminium chloride) and/or by EBPR. The latter is usually implemented by adding an anaerobic tank ahead the N-removal system (e.g. UCT, Bardenpho, A_2O, BCFS and BioDenipho systems). Thus the EBPR systems investigated included both biological N and P removal.

Selectors are not included in all of the plants studied. However, most anaerobic tanks in the Danish plants have several compartments, and the BioDenitro plants usually have a distribution channel in which return sludge and influent are mixed. Only few plants had "plugflow" basins.

6.4.4 Magnitude of bulking and foaming problems in Denmark, Greece and The Netherlands

Several surveys were carried out in the WWTPs of three countries, in which MLSS, SVI, DSVI, filament index (Eikelboom & Buijsen, 1983) and filamentous microorganisms identification were determined.

6.4.5 Settling properties in nutrient removal plants

In 60% of 20 Dutch plants upgrading to BNR increased the SVI. Similarly in Denmark, the number of plants with high SVIs also increased after introduction of nutrient removal (Kristensen et al., 1994; Andreasen & Sigvardsen, 1996)

The data from surveys of about 100 Danish nutrient removal plants, carried out in Jul. 1993, Dec. 1993 and Apr. 1994 showed a pronounced seasonal variation in SVI and DSVI at many plants (table 6.8). More than 20% of the plants surveyed had a DSVI > 150 mL/g in winter where only 14% exceeded this value in summer. In most cases the high SVI and DSVI values were associated with high filamentous microorganism levels.

Table 6.8. The seasonal variation in SVI, DSVI and Filament Index for approx. 100 Danish nutrient removal plants.

Data obtained from 3 screenings.

Survey	July 93	Dec 93	Apr 94
SVI Average (ml/g SS)	138	163	162
SVI > 150 mL/g (% of plants)	30	50	57
DSVI Average (ml/g SS)	115	124	130
DSVI > 150 mL/g (% of plants)	14	21	20
Filament Index (0-4)	2,4	2,5	2,7

A similar investigation was carried out in 18 Greek nutrient removal plants during the 5 year period from Nov. 1993 to Nov. 1998. SVI values recorded and microscopic observations made every 2 months showed a pronounced seasonal variation of both SVI and FI values (scale 0-5) (table 6.9). Most of the high SVI values were related to high FI values and 75 % of the high SVI were caused by *M. parvicella*. Usually, FI values > 2.5 resulted in SVI > 200 mL/g SS.

Table 6.9. The seasonal variation in SVI and FI (scale 0-5) for 18 Greek Nutrient removal plants. Data obtained from 30 screenings.

Screening	Summer	Winter
	(July-October)	(November-June)
SVI Average (ml/g)	160	200
SVI > 250 mL/g (% of samples)	-	30
FI Average	2	2.5
FI > 3.5 (% of samples)	-	28

Traditionally, the filament index has been used to assess the number of filaments in activated sludge. The ranking scale from 0 – 4 (Eikelboom index) is to some extent based on a personal judgement, wherefore a classification based on counting or real measurements may be an improvement (Sezgin et al., 1978).

For some of the Danish WWTPs the specific filament number (SFN) were therefore also determined together with the filament index. The SFN was determined by placing 50 μl of the sample under a 24x32 mm cover slip and then counting filaments extruding from the flocs, crossing a "counting transect line" from one side of the cover glass to the other (32 mm). Three transects were counted for each sample and the specific filament number was obtained by dividing the average filament count per transect by the MLSS in the sample (Jenkins et al. 1993).

The correlation between the Eikelboom index (EBI) and the SFN is shown in fig. 6.13; because the Eikelboom index is logarithmic in nature and has a maximum value of 4 (or 5/6), the data set is best fitted by a logarithmic expression. Especially at high filament indexes and SFN's there is a high scattering in the data set.

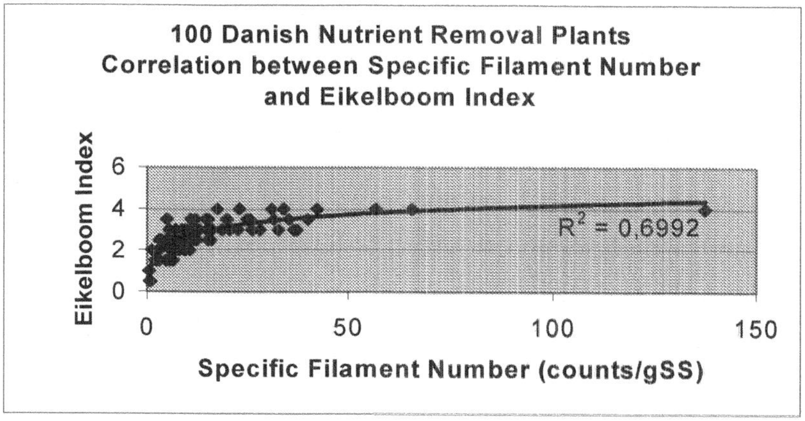

Fig. 6.13. Danish Nutrient Removal Plants Correlation between Specific Filament Number and Eikelboom Index

At each plant, a fairly clear (linear) correlation between DSVI and the specific filament number could often be observed: examples are presented for Herning (DK) WWTP (fig 6.14) and Metamorfosi pilot plant (fig. 6.15).

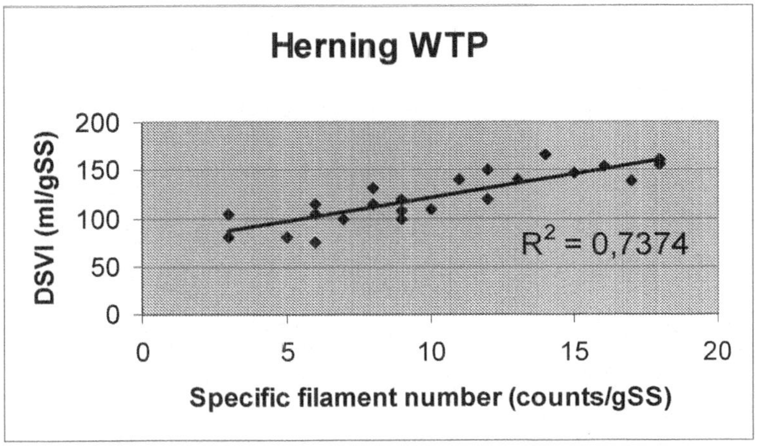

Fig. 6.14. Herning WWTP (DK). Correlation between Specific Filament Number and DSVI

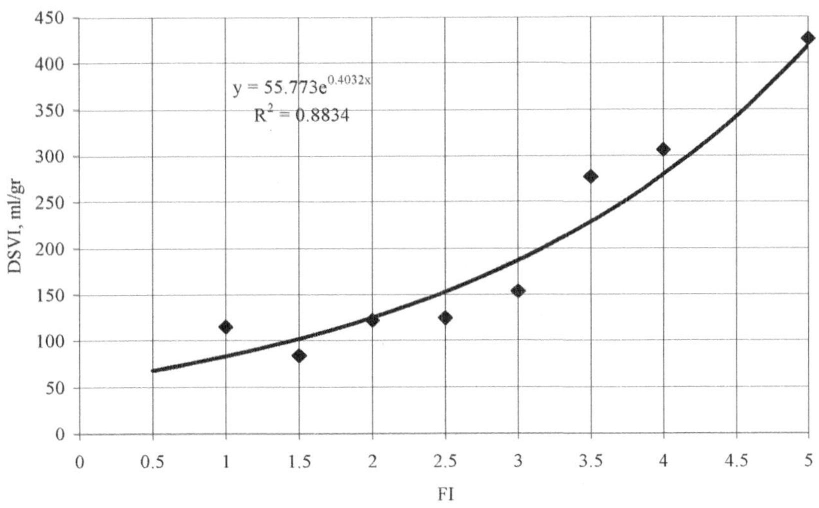

Fig. 6.15. Metamorfosi Pilot Plant (Greece). Correlation between Filament index (0-5) and DSVI (average value of all samples with same FI).

When comparing SFN and DSVI from different plants a weaker correlation and a higher scattering of the data set were found (fig. 6.16). This may be explained by the fact that factors such as the surface chemistry and density of the activated sludge flocs may influence the settling properties in a different way from plant to plant (Andreadakis, 1993). At two Danish treatment plants, aerobic granular sludge flocs were formed when the plants were operated at high MLSS and low DO concentrations, which could compensate for very high filament numbers.

It is also obvious that the various filamentous microorganisms may affect the DSVI in a different way. However, a very high DSVI (> 200 mL/g) was generally seen in connection with a high SFN (> 25 counts/gSS).

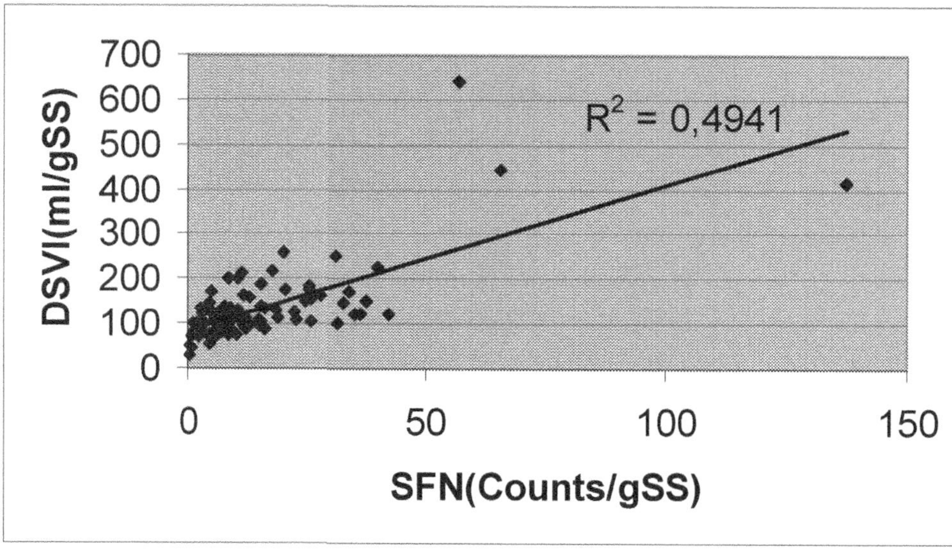

Fig. 6.16. Correlation between Specific Filament Number and DSVI for Danish nutrient removal plants

6.4.6 Effect of different process conditions on SVI and the size of the filamentous population

The influence of different process configurations on DSVI was investigated for Danish plants (Andreasen and Sigvardsen, 1996). EBPR plants had the best settling properties, especially in the absence of primary clarifiers. Because the FI for EBPR plants was often higher than for other plant configurations it was concluded that this was the result of a denser sludge. No clear DSVI differences were observed between predenitrification configurations and BioDenitro plants. With primary clarifiers BioDenitro configurations had the lowest SVI, but when no primary clarifiers were present they had the highest SVI.

Pilot plant studies in Greece during 1994-95 (Metamorfosi pilot plant, 400 m^3/d) showed that activated sludge from predenitrification system with plug flow anoxic and oxic zones had the best settling properties. This configuration controlled the growth of all filamentous microorganisms, including *M. parvicella*. The worst settling and the highest filamentous micro-organism growth were obtained with a completely mixed system with intermittent aeration for nitrification/denitrification. These findings were confirmed in laboratory scale experiments (Mamais et al., 1998).

The effect of selectors and other modes of operation were further investigated in a full-scale study at the Avedøre WWTP (DK). This oxidation ditch plant had experienced serious bulking due to *M. parvicella* after the introduction of simultaneous nitrification/denitrification. The FI increased from 1-

2 in the summer to 4 in the winter/spring; the DSVI increased from 125 mL/g to > 300 mL/g. The four configurations shown in table 6.10 were tested in parallel for more than 1 yr. (Jacobsen et al., 1998). Installation of an anoxic selector and the change from simultaneous to alternating nitrification/denitrification decreased considerably the filament abundance and DSVI, but changing to intermittent operation had no effect. After installing selectors in all plants and changing the mode of operation to alternating nitrification/denitrification the winter values of DSVI, FI and SFN decreased further. These results indicate that the effects of selectors and the mode of operation are cumulative. From the relative abundance of *M. parvicella* to Type 0041 it could be concluded that reduced FI and SFN were primarily caused by a reduction in *M. parvicella* number.

Table 6.10 Avedøre WWTP (DK). Effect of different modes of operation on DSVI, SFN and the relative abundance of *M. parvicella* and Type 0041. (Avg. data winter 94/95 and 97/98)

Line	Selector	N removal	DSVI	Filament index	SFN	Filaments	
			ml/g		counts/g SS	*M. parvicella*	Type 0041
						%	%
Winter 94/95:							
SDN	None	Simultaneous DN	258	3,5	64	90	10
Intermittent	None	On/off operation of aerators	255	3,5	63	81	19
Selector + SDN	Anoxic (approx. 30 min.)	Simultaneous DN	189	2,7	28	66	34
Bio Denitro	Momentaneous (< 1 min)	Alternating mode of operation	215	2,9	28	49	51
Winter 97/98:							
Selector + Bio Denitro	Momentaneous (< 1 min)	Alternating mode of operation	148	2,3	17	48	52

The surveys and pilot and full-scale studies indicate that filamentous population level in the various process configurations decreased in the following order:

N removal configurations:

- Simultaneous denitrification (highest)
- Intermittent nitrification/denitrification
- Alternating nitrification/denitrification (BioDenitro)
- Predenitrification (lowest)

P removal processes:

- EBPR (high filament number but dense flocs) (highest)
- Simultaneous precipitation with Fe salts
- Simultaneous precipitation with Al salts (lowest)

Flow pattern:

- Completely mixed activated sludge systems (CMAS) (highest)
- Selector + CMAS

- Plug flow pattern (tentatively the lowest but more examples needed to confirm this)

These rankings are only on the basis of relative number of filaments and not on the actual settling properties. As mentioned earlier, EBPR plants often have good settling properties despite a high number of filaments because denser sludge flocs. The same conclusion can be drawn from the recently upgrading of several Dutch WTPs to the BCFS process, where good settling properties are achieved due to dense sludge flocs. The usually high number of *M. parvicella* in EBPR plants may cause significant foaming problems.

Type 0041, which is mainly found in alternating nitrification/denitrification configurations may not decrease settling rate as much as *M. parvicella*, which was the dominant filamentous organism in the other configurations (Eikelboom et al, 1998). Andreasen and Sigvardsen (1996) found therefore also similar settling properties in terms of DSVI for alternating and predenitrification configurations under Danish conditions.

The addition of either iron or aluminium salts had no influence on the filamentous organisms, but often improved settling because of the production of dense sludge flocs.

6.4.7 Composition of the filamentous population

The most important filamentous microorganisms in BNR plants were by far *M. parvicella*, followed by Type 0041 (Denmark), Type 0092, *N. limicola*, Type 0803 and actinomycetes (Greece). Other filamentous organisms were seen in low numbers. The dominant filamentous organisms during spring 1994 are shown in table 6.11.

Table 6.11. Dominating filamentous species during spring 1994. Numbers indicate percentage of plants where a specific organism was present (from Eikelboom et al. 1998).

	The Netherlands					Denmark		Greece
Species	Oxidation Ditches	Carrousel	Schreiber Plants	Various Configurations	Bio-P	BioDenitro	Predeni-trification	Various Configurations
M. parvicella	58	66	44	67	74	28	43	55
Type 0041	6		11		32	38	20	
Type 0092	3				5	9		10
Type 1851			11		3			
Type 0803					5	3		
Type 0914					5			
N. limicola					16	3	10	
Actinomycetes								20
Type 021N		3	11			6		
Thiothrix sp.						6		
S. natans	3		11					
No of plants	31	32	9	21	19	32	30	11

Filamentous organisms which cause bulking in conventional activated sludge plants (*S. natans*, Type 021N, *Thiothrix*) are hardly ever found in nutrient removal plants. This may be because these filaments cannot denitrify and/or store energy for the substrate uptake in the anaerobic compartments of EBPR plants. Only when readily biodegradable organic substrate penetrates into the aerobic zones, do these filaments form a substantial part of the filamentous population in these types of plants. A situation where this occurs is when an industrial wastewater containing readily biodegradable substrate is fed to the aerobic zone.

M. parvicella populations varied seasonally with low numbers during the summer and autumn, an increase during the winter and a maximum in the early spring, explaining most of the SVI/DSVI variations observed. *N. limicola* showed a similar but less pronounced variation while Type 0092 showed the opposite pattern, with maximum levels during summer. Despite the seasonal differences in wastewater temperatures in Northern (8-16 °C) and Southern Europe (15-28 °C) it is interesting to see that the same seasonal pattern occurred.

6.4.8 Influence of wastewater composition, process configuration and operating parameters on filamentous microorganisms growth

Because many plants were investigated it was possible to obtain some knowledge about which parameters stimulate the growth of various filamentous microorganisms. A detailed description can be found in Eikelboom et al. (1998).

6.4.9 Effects of wastewater composition

In Danish WWTPs that received a high fraction of industrial wastewater from food processing industries, increased numbers of *N. limicola*, Type 021N and *Thiothrix* were present when some of the readily biodegradable organic matter passed to the aerobic zone. *M. parvicella* did not seem to grow well on Danish industrial wastewater.

M. parvicella is known mainly to utilise the lipid fraction of the wastewater (Slijkhuis, 1983; Andreasen & Nielsen, 1998), which forms a considerable part of the organic matter in municipal wastewater (Raunkjær et al. 1994). From the few data collected about lipids and LCFA in the wastewater, it was not possible to correlate the abundance of *M. parvicella* to a variation in the concentration of these compounds.

6.4.10 Effects of process configuration

Anaerobic conditions applied for Bio P removal stimulated the growth of *M. parvicella.* Applying presettling also increased the number of *M. parvicella*, which may be explained by the hydrolysis of the lipids in the primary settler.

Andreasen and Nielsen (1998, 1999) investigated the *in situ* physiology of *M. parvicella.* They found that *M. parvicella* did not utilise VFAs and other readily biodegradable COD under *in-situ* conditions; it was able to take up LCFAs under aerobic, anoxic and anaerobic conditions. Radioactive tracer studies with ^{33}P showed that it did not have a fast release and uptake of phosphate under alternating anaerobic and oxic conditions, but the uptake of LCFAs under anaerobic conditions showed that some energy storage compounds were present in the organism. As *M. parvicella* neither utilised VFAs nor had a rapid P release and uptake, it cannot be regarded as a normal Bio P organism. A competitive advantage towards the uptake of LCFA under anaerobic conditions is therefore the most likely explanation for the increased number of *M. parvicella* in Bio-P removal plants (Andreasen and Nielsen, 1999).

The general increase of *M parvicella* with the introduction of N removal could indicate that the organism is able to denitrify. However, the increase in sludge age, normally following the introduction of nutrient removal, may also be an important factor, as *M. parvicella* is believed to be a slow growing organism (Slijkhuis, 1983; Tandoi et al, 1998).

M. parvicella generally increased more with simultaneous and intermittent denitrification, than with predenitrification and alternating methods. This could indicate that *M. parvicella* may grow at a lower rate in anoxic rather than oxic conditions, where it is able to compete effectively at the low DO levels (K_{DO}: 0.016 mg/L; Slijkhuis, 1983) often present in plants with simultaneous denitrification. However, starvation experiments carried out with different electron acceptors (Andreasen and Nielsen, 1999) indicate that *M. parvicella* is able to utilize nitrate and nitrite as electron acceptors, whereas the pure culture study of Tandoi et al. (1998) only showed, that it was only able to utilise nitrate as an electron acceptor.

In BioDenitro systems, *M. parvicella* was found in much lower numbers than in other process configurations: Type 0041 took over instead. This could partly be explained by the selector effect introduced by the mixing of raw wastewater and return sludge in the distribution channels at BioDenitro plants, but also by the substrate gradients imposed by the alternating operation. Mixing of wastewater and return sludge mainly results in a removal of components in the particulate fraction (Eikelboom, 1987) and an adsorption of lipids/LCFA into the sludge flocs may explain the effect of mixing. A similar situation of reduced numbers of *M. parvicella* was observed in plug flow configurations, which besides the selector effect also introduced a substrate gradient throughout the treatment process.

The selector effect on *M. parvicella* has also been demonstrated in many Dutch oxidation ditches (Heide & Pasveer, 1974; Eikelboom, 1987), but the introduction of nutrient removal somehow reduces the selector effect. However, the investigation at the Avedøre WWTP (DK) showed that they may still have an important effect in nutrient removal plants.

6.4.11 Experience with control measures applied

Both specific and non-specific control measures have been applied in all 3 countries. It is important to recognise that the effect of the different control measures will vary from plant to plant, as filamentous bulking (growth rate of filaments) seldom depends on one single factor only. Therefore, conclusions based on single cases should be treated with great caution.

6.4.12 Non specific control measures

In Denmark, chlorine addition has been used to selectively kill the filamentous organisms. Although generally successful nitrification has been harmed when using high chlorine doses. Danish industrial and municipal nutrient treatment plants have long retention times, so chlorine addition to the aeration tanks is used. Doses of 2-10 kg Cl_2 /t SS d have been added batchwise to the aeration tank twice per day (50% in the morning and 50% in the afternoon) for a period of 3-10 days. The effect is normally seen within 1-2 weeks.

Traditionally, *M. parvicella* has been thought difficult to control by chlorination, but several Danish cases show that it can be controlled by the addition of 5-10 kg Cl_2 /t SS d. However, since nitrification is harmed by addition of more than 10 kg Cl_2 /t SS d, great precaution has to be taken and the effects on the nitrification evaluated prior to Cl_2 addition. As *M. parvicella* is a slow growing organism, the chlorination effect can last for months.

Chlorine addition may be environmentally questionable, and therefore other non-specific methods have been applied. In most cases, aluminium salts addition has reduced *M. parvicella* number considerably. For other filaments the effect of aluminium salts addition has been less significant. Aluminium has in general been added as polyaluminum chloride (e.g. PAX-14 from Kemira) to the RAS line at a daily dosage of 1-3 kg Al/t SS.d. An example of the effect of aluminium addition at Åby WWTP (DK) is shown in Fig. 6.17. After 2 months dosing to the return sludge line, the DSVI was reduced from approximately 175 to below 100 mL/g. SFN simultaneously decreased from 20 to 10 counts/g SS in the test line, while no reduction was observed in the reference line. The applied dosage was 1.5 kg Al_2 /t SS d, reduced to 0.7 kg Al/t SS d after 1 month; higher aluminium dosage may have a negative effect on the nitrification (Jacobsen et al., 1998). Recent experiences from The Netherlands indicate a dosage according to the following formula:

Aluminum dosage (kg Al/t SS d)= 60/sludge age

therefore this solution will bee costly at short sludge ages. No inhibition of the nitrification was observed at this dosing rate. In general the addition of Al will also reduce the foaming problems caused by *M. parvicella*.

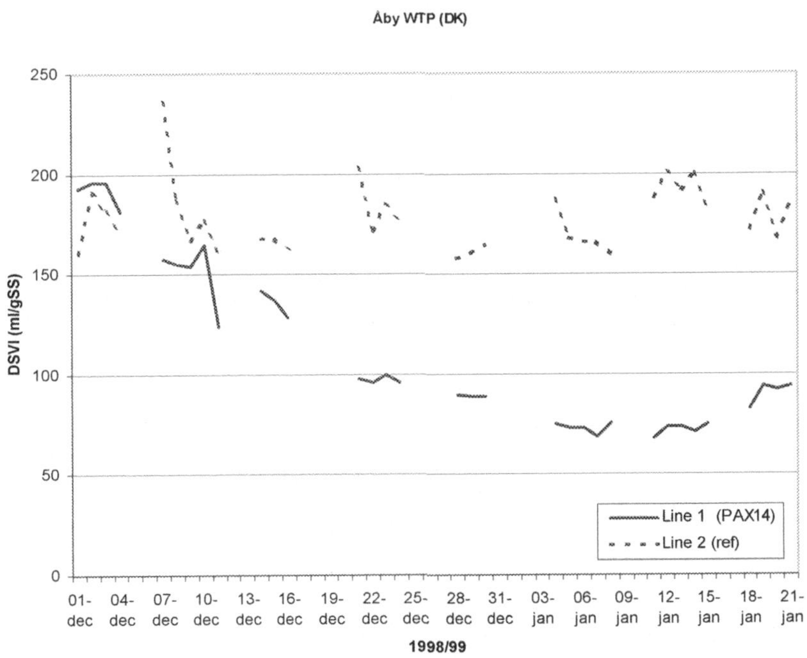

Fig. 6.17 WWTP (DK). Effect of the dosing of polyaluminum chloride (PAX 14-Kemira) on the DSVI. Dosing started on 1st Dec. 1998.

The addition of talc has been used in The Netherlands and Denmark to control bulking at industrial plants. Talc has been added to the aeration tanks at an initial dose of 80-100% of the biomass inventory. Further additions of talc are made at the same mass ratio in proportion to the daily sludge production.

The effect of talc addition is almost instantaneous and nitrification is not inhibited (Eikelboom and Grovenstein, 1998). Talc does not increase the waste sludge volume because the higher solids concentration in the dewatered sludge compensates for the increased sludge production. However it does not increase the solids loading on the secondary clarifier.

6.4.13 Specific control measures

Selectors:
During the last 2 decades, selectors have been established as a normal part of most treatment plants. Industrial plants are seldom built without them.

For filaments found in conventional activated sludge plants (e.g. *S. natans*, Type 1701, Type 021N and *Thiothrix*) the most pronounced effect is obtained with anaerobic and anoxic selectors (Wanner, 1994), because these filaments have only limited activity under these conditions. The EU project verified that most of these filaments could be controlled by the anoxic or anaerobic process conditions implemented in connection with BNR, if most of the readily biodegradable substrate was removed in the selector. For these filaments, the anoxic and anaerobic selectors sizing should be based on traditional sizing techniques for BNR plants (Wanner, 1994; Jenkins et al. 1993).

At many industrial plants there may be insufficient N and P in the wastewater to ensure an efficient removal of the readily biodegradable COD in the anoxic or anaerobic selectors (e.g. plants treating paper industry wastewater). In these cases, an aerobic selector could be installed, but the general experience is that the kinetic selection pressure imposed by aerobic selectors is weaker than the metabolic selection imposed by anaerobic and anoxic selectors. An alternative is to add nitrate to the water. This is efficient but expensive compared to the installment of an anoxic selector. In a paper industry bulking problems were solved by changing the plant from a single-stage system to a two-stage system, operated with F/M ratios of approximately 1 and 0.2 kg COD/kg SS·d respectively.

The filamentous populations at industrial treatment plants may be much more complex than at municipal plants. New undescribed filamentous types can be found in these plants. Oligonucleotide probes and the recently developed radiographic technique for in-situ characterisation of the physiology of filamentous microorganisms may be a valuable tool for establishment of the best control strategy (Eikelboom and Geurkink, 2002; Andreasen and Nielsen, 1997; Nielsen et al., 1998). This method makes it possible to investigate substrate preferences and uptake capacity under different process conditions without isolation and cultivation. Hence, the effect of the metabolic selection could be investigated prior to the implementation.

M. parvicella:
M. parvicella is the most common filamentous microorganism in BNR plants, and therefore it deserves special attention. The proliferation of this filament in BNR plants shows that metabolic selection (anaerobic and anoxic conditions) does not control this filament. In fact the initial anaerobic conditions used for Bio-P removal seem to stimulate the growth of this microorganism. This disagrees with the first experiences with anaerobic compartments for Bio-P removal, which

were shown to control filamentous bulking (Tracy, 1973). However these investigations were collected mostly from highly loaded systems (MCRT<5 d), where *M. parvicella* does not proliferate. The MCRT used for BNR in Europe is often > 20 d, where it is more likely that substrate availability rather than maximum growth rate, determines *M. parvicella* growth rate. The effect of MCRT on the growth of *M. parvicella* was investigated by Knoop and Kunst (1998), who showed that unnecessary long MCRT should be avoided in BNR plants.

The mixing of raw wastewater and return sludge (instantaneous selector) reduce the growth of *M. parvicella* in BNR plants, but may not always control it effectively.

At most Danish EBPR plants the anaerobic zone is divided in 3-4 compartments to impose a substrate gradient through the anaerobic step and optimise EBPR. This gradient has not been enough to control the growth of *M. parvicella*. Since *M. parvicella* seems to have the capability to store LCFA released under anaerobic conditions, the retention time in the anaerobic compartment may be important and should be reduced to a minimum to obtain stable Bio P removal. The installation of a secondary selector after the anaerobic stage has improved the situation at some Dutch treatment plants (van Loosdrecht et al., 1998).

Lipids and LCFA are fairly insoluble in water, so that the effect of selectors in controlling *M. parvicella* may be better understood on the basis of adsorption of lipids/LCFA into the sludge flocs than from a change in the concentration presented to the microorganisms. A substrate gradient imposed by alternating feeding or plug flow seems to have a positive effect on *M. parvicella* control. Should this effect be explained by a better adsorption of LCFA to flocformers, or by flocformers improved uptake kinetics throughout the treatment process, has still to be elucidated.

The choice between completely mixed activated sludge systems (CMAS) and plugflow systems is a trade-off between many factors (e.g. flexibility, DO control, construction and operating costs). It seems that selectors can be used to remedy most of the bulking problems in CMAS systems.

For nitrogen removal processes, experience from the 3 countries indicates that simultaneous and intermittent denitrification should be avoided. However full-scale experience from Avedøre WWTP showed that the installation of an anoxic selector might improve the situation in plants with simultaneous denitrification (Jacobsen et al., 1997). A more pronounced positive effect may be expected using separate nitrification and denitrification together with a large substrate gradient.

An improved understanding of the hydrolysis and degradation of lipids under different process conditions is needed before the control of *M. parvicella* can be optimised.

Other specific control measures:

A constant, low DO (<0,5-1 mg/L O_2) in the aeration tank has been shown at some Danish BNR plants to stimulate *M. parvicella* proliferation. This is consistent with earlier Dutch experiences (Slijkhuis and Deinema, 1988).

6.4.14 Conclusions

Experience in Denmark, Greece and The Netherlands shows that filamentous bulking may be increased in BNR plants due to *M. parvicella* proliferation. Guidelines have been presented to allow acceptable sludge settling characteristics in most BNR plants.

Acknowledgement
The European Commission under the environment Programme supported the EU-study. The Danish Academy of Sciences supported part of the work under an industrial PhD-study. The staff at the involved treatment plants are acknowledged for their effort in collecting and providing data for the project.

6.5 FRANCE (R. PUJOL)

6.5.1 General situation

In France, activated sludge plants account for approximately 80% of all facilities (15,000 in total).

A majority of these (about 90%) use extended aeration with one or several aeration tanks, depending on the size of the plant. Many of these plants suffer from bulking and foaming problems. Reliable remedial actions to face every particular problem require a consistent approach, based on three steps:

- the precise characterization of the microorganisms involved by using the available identification keys
- the comprehensive study of the plants operation, (facility limits, origins and aggravating factors for filamentous growth)
- the selection of suitable technical solutions.

This chapter presents the results of two french surveys, relates filamentous microorganisms to plant operation parameters and gives two solutions successfully adopted for filamentous growth control, based on the installation of a "contact zone" (kind of selector with only one compartment) and chlorination.

6.5.2 Bulking and foaming characterization

In France, bulking sludge is characterized by a DSVI > 200 mL/g associated with the presence of a large number of filamentous micro-organisms.

Either in addition with bulking sludge or not, stable biological foams can be observed at the surface of many plants. The stable biological foam is described as a system comprising three components:

Specific filamentous microorganisms (nocardioforms, *M. parvicella).* + **Gas bubbles** + **Floc particles** = **FOAMING SLUDGE**

6.5.3 Bulking and foaming situation

Most of the information presented here is from 2 surveys of 63 and 75 % of the French plants (Pujol and Canler 1989; Pujol *et al.* 1991).

These surveys showed that:

- about 25 % of the plants have SVI > 200 mL/g at least once per year,
- about 20% of the plants have occasional or permanent foaming.

The F/M ratios of the plants are shown in table 6.12.

These biological problems develop mainly in extended aeration plants regardless of treatment capacity, geographical location or aeration device (surface aerators, bubbles aeration etc.); 76% of the plants treat municipal wastewater, but when industrial wastewaters (dairies, slaughterhouses, etc) are treated the problems are much greater.

Table 6.12. F/M ratio of bulking and foaming in activated sludge plants

Sludge F/M	KgBOD/kgMLVSS.d	% of case	
		bulking	foaming
Extended aeration	< 0.1	79	87
Low	< 0.2	12	10
Middle	0.2-0.5	7	2
High	> 0.5	2	1

Scum appears to be cyclic or repetitive in 50% of the foaming cases, occurring mostly in winter and spring.

6.5.4 Dominant filamentous microorganisms in bulking and foaming sludges

Microscopic observations have been carried out on a large sample of sludges, according the identification criteria of specific Manuals (Eikelboom's and Van Buijsen, 1983; Jenkins *et al* 1986).

Results of filamentous microorganisms abundances are summarized in Table 6.13.

Table 6.13. Main filaments in bulking and foaming sludge (low F/M ratio)

Activated sludge		Foam	
Filamentous microorganism	Rank	*Filamentous microorganism*	associated filaments
Type 0041	1	*M. parvicella*	
Type 0092	2	Nocardioform	type 0041
M. parvicella	3		type 0675
Thiothrix	4		type 0092
Type 021N	5		type 021N

All these filaments are commonly observed in low F/M activated sludges where substrate concentrations are weak in the mixed liquor, and with high sludge retention time (>20 days).

6.5.5 Filamentous microorganisms and relationship with plant operation

M. parvicella was the filamentous microorganisms most often observed in the analyzed samples (bulking + foaming), this filament has been emphasized by several authors (Slijkhuis and Deinema, 1982; Eikelboom and Van Buijsen, 1983; Blackbeard *et al.*, 1986). Very high SVI (400-500 mL/g) can be reached with *M. parvicella.*

The main causes for *M. parvicella* growth are low F/M, wastewater characteristics (influent septicity, sludge treatment returns, high VFA concentrations, surfactants, etc.).

Process features that worsen the foaming problem are surface trapping (due to physical barriers or hydrodynamics effects), denitrification in the bottom of the secondary clarifier, scarse aeration conditions, etc.

Type 0041 frequently found in activated sludge bulking is also associated with foaming problems. Its levels are often lower in foam than in mixed liquor so that Type 0041 exhibits less flotation ability (its cell wall nature is probably less hydrophobic). In France, Type 0041 appears to be the commonest filamentous microorganisms in low F/M bulking sludge, able to increase SVI up to 500 mL/g. The causes for its development are related to nutritional deficiencies worsened by some limitation in aeration or mixing conditions. Nocardioforms, that have a quite complex taxonomy (Lemmer and Kroppendstedt, 1984; Kahn and Forster, 1988) were prevalent in only 14 % of foaming cases. Because of their morphology and heterogeneous spatial distribution in activated sludge, Nocardioforms do not induce notable deterioration of sludge settleability (SVI < 200 mL/g).

Regarding Type 0092, the number of problems attributed to this bacterium is limited.

Thiothrix and Type 021 N filaments have been identified in some samples, due to wastewater septicity or nutrient deficiency.

6.5.6 Successful solutions for bulking and foaming control

Precise identification of the microorganisms involved and knowledge of their ecophysiology, remains the basis for choosing reliable solutions for the filamentous microorganisms proliferation control. Essential is also the comprehensive study of the plant's operation (facility limits and filamentous organisms origin and aggravating factors).

Activated sludge foaming requires even more attention because the problem is worsening with time. Initially the foam layer is thin, present only in some plant surfaces (aeration corners, desorption tank, etc.). In such a situation, it is important to monitor foam evolution correlated with the SVI (secure threshold value: 150 mL/g).

The first actions should be to control the aeration/mixing conditions, sludge return septicity, and sludge concentrations, mechanical extraction of floating sludge without recycling back in the aeration basin.

As a preventive solution (or curative when the plant is already in operation), suited for low F/M ratio activated sludge plant, a contact zone (Pujol, 1984) can be easily implemented. It is a sort of selector (Chudoba *et al.*, 1973) consisting of one small reactor installed upstream the aeration tank and operating with a mixture of raw wastewater and recirculated sludge. This reactor can be installed in a new plant or in an existing plant. Many plants have been implemented with success as it is shown in chapter 6.5.7.

When time is limited and/or acute biological problems occur, and/or traditional solutions are not applicable, it can be necessary to use curative solution by toxic compounds such as chlorine (chapter 6.5.8). If correctly utilised, it remains the most economical non-specific method for controlling filamentous organisms growth (Jenkins *et al,* 1993). This solution has to be adopted for limited time, due to possible negative effect on the environment (such as AOX formation).

6.5.7 Contact zone as preventive solution

Principle: The contact zone technique (Fig 6.18) entails the establishment of a zone containing high concentration of readily biodegradable substrate, which stimulates substrate uptake and storage capacities of floc-forming organisms.

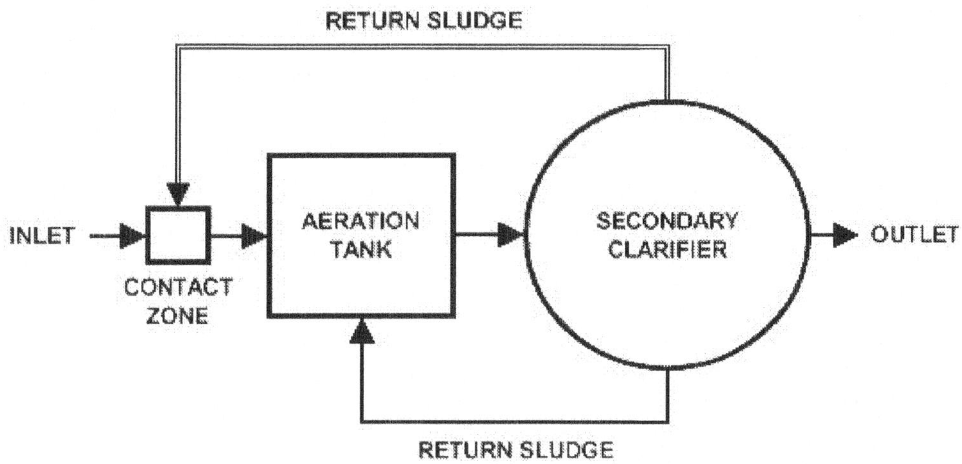

Fig. 6.18. The contact zone flowsheet

This principle is based on earlier works carried out by Chudoba *et al.*(1973), Eikelboom (1982), and Chiesa and Irvine (1985).

In terms of plant design, the establishment of a contact zone requires installation of a small tank upstream the aeration tank, in which sludge (generally return activated sludge) and primary effluent are combined in a given proportion.

Application of this technique is reserved for environments suffering from a slight nutrient deficiency. The optimum scope of application is reserved for low-F/M activated sludge plants handling municipal or industrial (particularly food industry) wastewater. Contact time in the contact zone is about 10 minutes, and the loading in the zone is equivalent to 100 mg of readily biodegradable COD per gram of MLSS (Pujol and Canler,1992). Design values are based on the peak loading rates. Mixing is mandatory, aeration optional (oxygen concentration in the zone is zero or almost zero at any time).

Results achieved using contact zones. Twelve extended aeration facilities (with capacities ranging from 600 to 30,000 PE) some of which had primary clarification, were equipped with a contact zone. The duration and degree of monitoring varied from one facility to another, but was at least 6 months in all cases (Pujol and Canler 1994).

Table 6.14 summarises the facility characteristics and the results.

Table 6.14: Experimental results from contact zones

PLANT	CAPACITY	WASTEWATER	CONTACT ZONE		EFFICIENCY		FILAMENT
	PE	COD (mg/L)	Volume (m^3)	Aeration	SVI*B/A	FOAMS**	
1	30000	600	160	yes	400/100		Type 0092
2	600	800	2	no	800/150		Type 0041 *M. parvicella*
3	10000	850	25	yes	800/150	+ -	*M. parvicella* Type 0041
4	1500	510	6	yes	500/200	+	*M. parvicella* *Nocardia sp*
5	3000	760	6	yes	250/250	+	*M. parvicella*
6	26000	340	125	yes	400/100	+	*M. parvicella* Type 0041
7	1000	450	2	no	250/170	-	*M. parvicella*
8	1500	900	10	yes	150/100	+	*M. parvicella* *Nocardia sp*
9	15000	925	19	yes	350/200	-	*M. parvicella*
10	6000	480	11	no	300/200	+ -	*Nostocïda I*
11	14000	800	30	yes	300/150	+	*M. parvicella*
12	12000	380	60	no	300/200	+ -	*M. parvicella*

* B: before, A: after implementation contact zone; ** +: positive effect, -: negative effect; + -: variable effect

Wastewater were primarily of domestic origin, with average COD levels ranging from 340 to 925 mg/L. The dominant filamentous microorganisms were typical of extended aeration systems. In plants 4 through 12 show *M. parvicella* was the dominant filament (in both the foam and the sludge).

In all cases, the contact zones were small (< 3%) in comparison with the aeration volume. Eight of the zones were equipped with aeration (usually an air blower in the grease removal tank), which improved mixing and promoted contact between sludge and wastewater.

System performance was measured by two criteria:

- SVI which showed a marked decrease for 11 out of 12 plants
- Presence of biological foam suppressed in 75 % of the plants, including total removal in over 50% of the sampling. In two plants (4 and 6), the contact zone did not reduce foaming, but did have a positive impact on the SVI.

6.5.7.1 Key points of contact zone design

• *Loading*. Contact zone loadings were on the order of 100 mg COD/g SS, and ranged up to 300 mg COD/g SS. The contact zones were often installed in the grease removal units for the sake of economy and expediency. As a result, the loadings do not always comply strictly with typical sizing calculations. But organic loadings were much lower than those indicated in the literature (Wanner 1994; Albertson 1991, Casey *et al.* 1992).

• *Contact time:* Theoretically, the sludge/wastewater contact time is setted at 10 minutes (peak flow rate plus return sludge flow rate). In the plants tested, the wastewater/sludge mixture remained about 1/2 h in the contact zone on average. Previous testing showed that biosorption could occur with times of this order of magnitude. If the contact time is too long, floc-forming and filamentous organisms have equal chances to use the available substrate in the contact zone.

These observations support the hypothesis advanced by Chiesa and Irvine (1985), that filamentous bacteria in low F/M activated sludge plants have a lower substrate uptake rate than floc forming bacteria. At the opposite end of the spectrum, insufficient contact times are detrimental to process efficiency; loading rates are too often high and the relevant sludge volume too low.

• *Aeration - mixing:* Is aeration in the contact zone necessary ? Studies have indicated that foaming is better controlled with one aerated zone: in a four "non-aerated contact zones" (Table 6.14), the results were not conclusive. This might be taken as proof that aeration is essential in the contact zone. However, it appeared that for Plant 9 (aerated contact zone), aeration did not reduce *M. parvicella* foam.

An examination of DO levels in the contact zones showed consistently values equal to or approaching zero (whether or not the zones were aerated). Even though oxygen presence is not absolutely necessary, in case of septic wastewater it is recommended to have an aerated contact zone.

It was essential for wastewater and sludge to be introduced into the contact zone simultaneously, since mixing undoubtedly plays a crucial role. Further improvement in the results may have been observed if such mixing criteria had existed (consistently particularly in Plants 7, 10, and 12).

Progressive efficiency: The results achieved show that the contact zone is not immediately effective (Fig. 6.19). The SVI is decreased gradually: the higher the initial SVI, the faster the decrease occurs. Generally, 2-3 sludge retention time were required before the situation became stable. The use of a contact zone is an effective option, but requires time to show full effect.

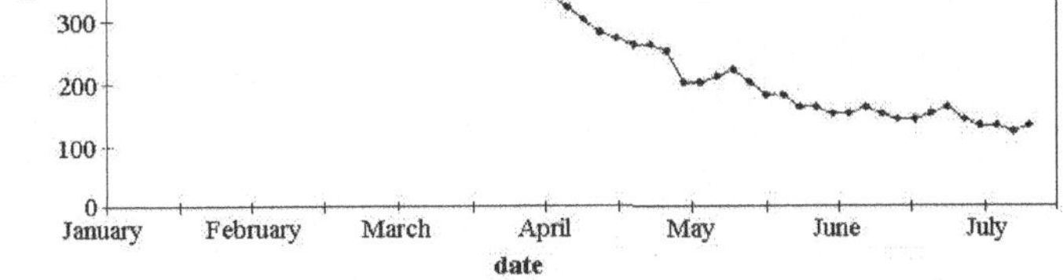

Fig. 6.19. SVI evolution after implementation of a contact zone in plant n° 3

Plant operating parameters: It goes without saying that plant operating parameters strongly influence the bulking and foaming. It is important to provide sufficient aeration to prevent *M. parvicella* growth.

It is also important to manage sludge production by proper wasting. Unfortunately sludge wasting is not always under plant operator control, for several reasons (such as bans on agricultural re-use, limited plant sludge treatment capacity, associated costs, etc.): this have consequences that worsen the problems (high SS concentrations in the tank, DO reduction, septic return sludge, etc.).

Conclusion: For the studied plants, the SVI reduction and foaming intensity confirmed the effectiveness of contact zones: SVI improved in 91% of the cases and foam was controlled in 75% of the cases. These results were achieved with contact zone loading rate of approx. 100 mg dissolved COD/g SS, and a minimum contact time of 10 minutes (during peak conditions, including sludge return flow).

Mixing in the contact zone and simultaneous introduction of wastewater and sludge were essential for process efficiency. Improving the situation took generally 2-3 times the sludge age. This can arrive to several months in some low F/M facilities.

The solution is not universal. Bulking related to *Beggiatoa* cannot be remedied by implementation of a contact zone, neither can the effects of insufficient aeration.

The technique is relatively low-cost, easy to install (in new or existing facilities) and entails few additional operating constraints. Rational management of aeration and sludge production will ensure effective use of this process.

6.5.8 Control of bulking by Cl_2 addition

Plant presentation and situation: The plant, located near Paris, treated wastewater of 1.2 10^6 PE (300 000 m^3d). It was a low F/M activated sludge plant. The plant consisted of four lines preceded by a primary settling tank. The two first lines have been operated since 1987 and the others since 1992. This plant achieves simultaneously nitrification and partial denitrification by initial anoxic zones in the aeration tank (HRT: 8 hours). Periodically, the plant showed high SVI associated with foams and *M. parvicella*, sometimes *Nocardia* (table 6.15).

Table 6.15. SVI evolution before application of the control strategy

Month	Sludge volume index (ml/g)				effluent S.S
	line 1	line 2	line 3	line 4	mg/L
September 1996	76	69	82	78	14
October	76	64	89	104	15
November	62	64	144	178	28
December	71	74	219	240	57
January 1997	97	177	242	298	61
February	167	212	286	353	69
March	139	181	218	270	52
April	102	160	177	172	23
May	67	80	151	103	27
June	78	70	91	85	15

(SS effluent standard=30mg/L)

The phenomenon was aggravated by a long HRT in anoxic zone (> 2 h), return sludge from digested storage tanks, and wastewater septicity. Usually the main difficulties were observed during wintertime. Despite of modifying operating conditions the problem was not solved.

Chlorination procedure: Effect of chlorination is well reported in the literature (Jenkins et al., 1986) for controlling filaments growth. Some warning criteria and specific actions have been adopted:

- regular microscopic observation of sludge and scum samples to detect in advance filaments growth
- SVI monitoring; when SVI was >150 mL/g, short time chlorination started with a maximum dosage of 2 kg Cl_2/t SS.d
- supervision of the aeration needs, actions on the blowers and optimisation of air distribution in the aeration tanks
- regular analysis (several times per day) of NH_4^+, effluent concentrations to check that it remained below 2 mg/L every time
- observation of foaming accumulation in the plant and its extraction when increased

Results: Table 6.16 and figure 6.20 present the results of RAS chlorination during wintertime (Moryoussef and Trousselle, 1998). The duration of chlorination is never longer than 48 h; the effect of chlorine on the SVI was immediate and the SVI target of150 mL/g was never exceeded. Table 6.16 shows the effluent quality discharged into Seine river since the application of the chlorination procedure.

Table 6.16. SVI evolution after application of the control strategy

Month	Sludge volume index (ml/g)				Effluent SS
	line 1	line 2	line 3	line 4	mg/L
July 1997	61	61	85	86	14
August	53	43	62	66	18
September	64	59	81	96	13
October	63	60	112	128	15
November	72	66	127	127	23
December	67	60	125	135	27
January 1998	98	71	128	123	23
February	104	98	127	125	14
March	72	72	125	129	25
April	59	67	132	98	23
May	62	64	111	100	18

(SS effluent lineat standard = 30mg/L)

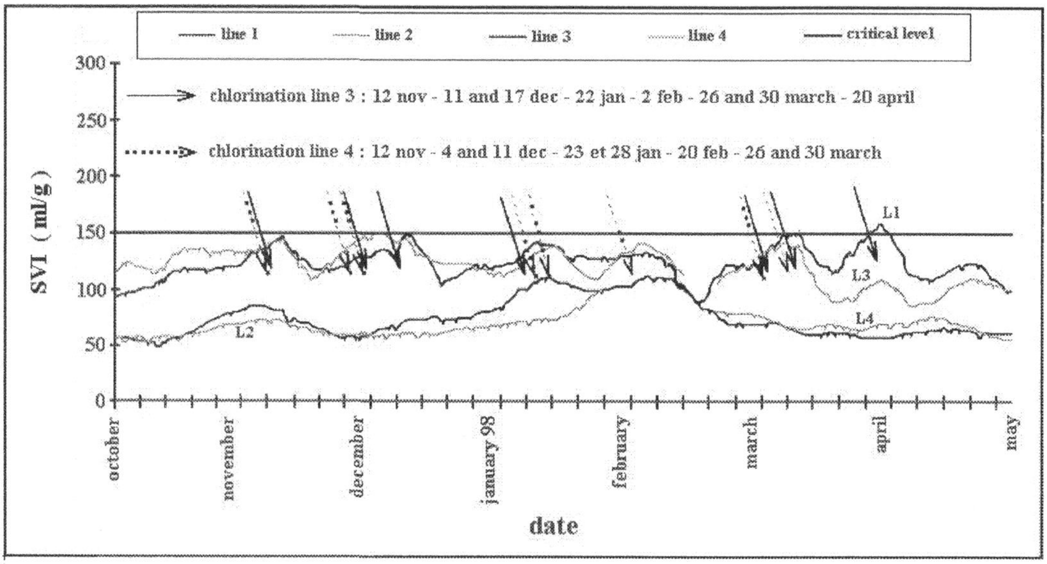

Fig. 6.20. SVI values during application of control strategy (Moryoussef and Trousselle, 1998)

Conclusion: Chlorination is always efficient when it is well managed. Chlorination effects are temporary and cannot eliminate the causes of filamentous growth. chlorine dosage has to be suited for each case, applying the lower efficient dose.

6.6 GERMANY/AUSTRIA (H. LEMMER AND N. MATSCHÉ)

6.6.1 General situation

The most striking difference in wastewater treatment since the 1990s, as compared to former performance, is the introduction of nutrient removal in order to comply with German directives (Abwasserverordnung AbwV) from 1994 and the more restrictive version from 1997. They fix limit values for total nitrogen concentrations of 18 and 13 mg/L, respectively, for municipal secondary effluents of WWTP exceeding 10^5 PE.

6.6.2 Bulking and scum in municipal and industrial wastewater treatment plants

Therefore the most obvious difference between the situation end of the 1980s and now is a strong shift from municipal carbonaceous substrate removal plants to nitrification/denitrification (N/DN) and biological phosphorus removal (B-P) plants. This shift is established since the mid 1990s as is shown in Fig. 6.21 where sludge separation problems, F/M ratio, and type of nutrient removal in the WWTP surveys of 1989 and 1996 are compared (Lind and Lemmer 1998). As far as industrial WWTP is concerned, we still face high load conditions in many cases with carbonaceous surplus and respective shortages in both oxygen and nitrogen/phosphorus supply.

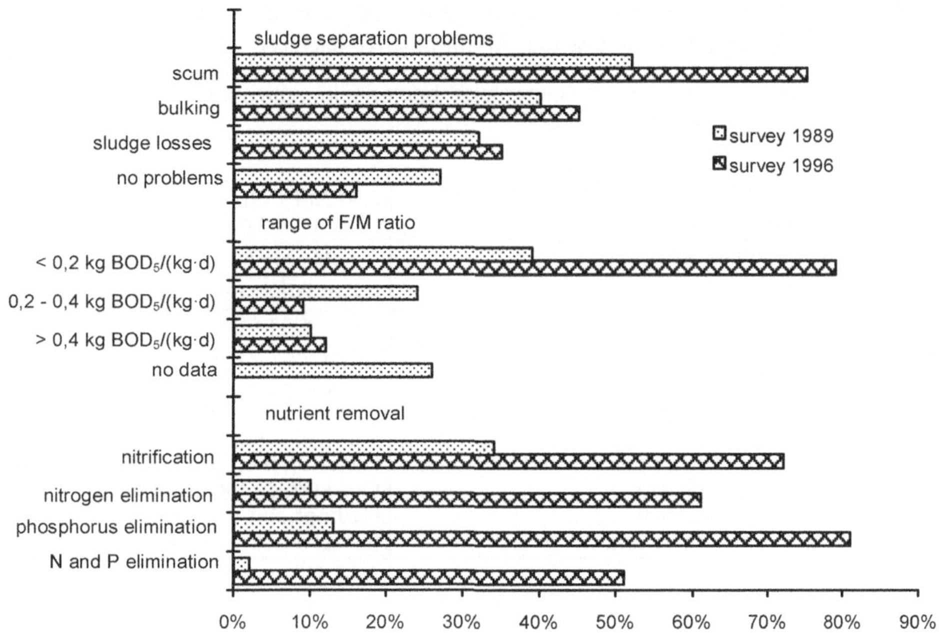

Fig. 6.21. Sludge separation problems, F/M ratio, and type of nutrient removal in wwtp surveys of 1989 and 1996 (from Lind and Lemmer 1998)

The situation in Austria is quite similar, and the present chapter contains some Austrian case studies on the most common problems. The problems in Switzerland are somewhat different because of different legislative directives.

With respect of filamentous activated sludge bacteria being involved we presently observe in Germany a marked predominance of scum promoting bacteria as compared to bulking microorganisms, in particular in municipal WWTP at F/M ratios below 0.2 kg BOD_5/(kg MLSS·d). *M. parvicella* was the dominant filament, followed by nocardioform actinomycetes and Type 0092 (Müller *et al.* 2004). Various other morphotypes such as *Nostocoida limicola* or Types 0041/0675 and 1851 played a minor role in municipal scum problems (Lemmer *et al.* 2004). Industrial WWTPS with generally higher loads are still suffering from bulking problems, due in particular to *Sphaerotilus* spp, Type 021N, or specific other industrial filamentous bacteria being involved (Snaidr *et al.* 2002).

6.6.3 Biology and specific troubleshooting measures of filamentous bacteria

At present three types of sludge separation problems are experienced (Table 6.17). Sludge bulking due to sulfur bacteria, sludge bulking due to high load Gram-negative filamentous bacteria, and scum formation due to low load Gram-positive bacteria.

Table 6.17. Physiological groups of filamentous bacteria in activated sludge with important selection factors (after Lemmer and Lind 2000)

SULFUR BACTERIA	GRAM-NEGATIVE BACTERIA HIGH F/M RATIO	GRAM-POSITIVE BACTERIA LOW F/M RATIO
Thiothrix spp.	*Sphaerotilus* spp.	*"M. parvicella*
Beggiatoa spp.	Type 1701	nocardioform actinomycetes
Type 0914	*Haliscomenobacter hydrossis*	Type 0041 / 0675
Type 021N	Type 0961	*"Nostocoida limicola" I / II / III*
	Type 021N	Type 1851
	Type 1863	
Reduced sulfur compounds septic conditions	F/M > 0,15kg BOD_5/ (kg MLSS ·d) N/P- deficiency O_2-deficiency complete mixing	F/M < 0,15 kg BOD_5 / (kg MLSS ·d) Cell surface hydrophobicity Hydrophobic substrate Surface-active substances

In 1998 the task group ATV-DVWK KA 6.1 "sludge bulking, scum, biological additives" published information on the specific "handling" of these organisms (Lemmer *et al.* 1998). Many of these hints are still valid: a summary of these follows.

6.6.4 Sulfur bacteria

Sulfur bacteria (*Thiothrix*, *Beggiatoa* or Type 0914) (see Table 6.17) can utilize reduced sulfur compounds as an energy source by oxidizing it to elemental sulfur and eventually to sulfate. In the case of municipal plants anaerobic conditions producing sulfides might be due to long retention times in the sewer system, or to sludge deposits in the aeration tanks. Under these conditions filamentous sulfur bacteria are able to increase in number so that separation problems can arise. These bacteria are found in an F/M range of 0.15 to 0.6 kg BOD_5/(kg MLSS·d) in both industrial and municipal plants with septic wastewater containing reduced sulfur compounds and short-chain fatty acids. The oxygen content in the primary effluent most often is below 0.1 mg/L O_2. At F/M values below 0.12 kg BOD_5/(kg MLSS·d) as applied in municipal N/DN and Bio-P plants filamentous sulfur bacteria are rare.

Control measures aim at decreasing the content of both reduced sulfur compounds and readily biodegradable substrates in the primary effluent. This can be achieved by a precipitation step preceding the primary clarifier or by aerating the primary effluent. Optimizing aeration and mixing performance may help reduce septic conditions.

6.6.5 Gram-negative filamentous bacteria in highly load plants

The second group of filamentous bacteria causes bulking in non-nitrifying industrial WWTPS with F/M ratios exceeding 0.2 kg BOD_5/(kg MLSS·d) and/or in WWTPS subject to frequent shock loading. This group comprises *Sphaerotilus* sp., *Haliscomenobacter hydrossis*, and Types 021N, 1701, and 1863 (see Table 6.17). These organisms are also selected by nutrient-deficiency conditions due to high concentrations of carbonaceous substrate and/or N- and P-limitation, or also low oxygen supply. Completely mixed systems with an even substrate distribution favored these organisms in German WWTPS (Sarfert *et al.* 1988). Therefore introducing a substrate gradient by cascades or plug flow conditions (Chudoba et al 1973) provided a promising means for selective troubleshooting. A high load premixing zone (aerobic selector) is able to prevent bulking events caused by these organisms. Aerobic selector design experience has been presented by Kroiß *et al.* (1997) and Prendl (1997).

6.6.6 Experience with selectors for industrial wastes in Austria

During investigations on treatment plants at sugar mills and a paper mill it was demonstrated that activated sludge bulking caused by filamentous bacteria Type 021N, Type 0041 and Type 1701 could be suppressed by an "optimum aerobic selector system". Bulking control was only reliable if the dissolved readily biodegradable substrate was completely removed in the selector. The composition of the readily biodegradable substrate (fresh or acidified) influenced substrate removed rates and oxygen requirements at selector. Selectors are designed so that the time for proper substrate storage and sufficient oxygenation capacity is granted. Storage capacity depends on the sludge age. Also at sludge ages below 5 d, the storage capacity cannot be recovered in the aeration tank leading to the loss of the selector effect.

6.6.6.1 Dimensioning and construction of an aerobic selector

Bulking sludge can only be successfully prevented with an aerobic selector if the readily biodegradable soluble substrate is rapidly and almost completely removed in the selector. For this to occur the selector retention time must be long enough. The required selector volume can be calculated according to equation 1.

$$V_{SEL} = \frac{Q_{INF} \cdot S_{SINF}}{MLVSS_{SEL} \cdot q_{min(SS.OUR)}} \qquad (1)$$

V_{SEL}	selector volume (m^3)
Q_{INF}	inflow (m^3/h)
$S_{S\ INF}$	influent soluble readily degradable substrate (g/m^3)
$MLVSS_{SEl}$	mixed liquor volatile suspended solids concentration in selector (kg/m^3)
$q_{min\ (SS,\ OUR)}$	minimum specific substrate elimination rate which can be expected as a function of the concentration of the readily degradable substrate and the oxygenation capacity in the selector (kg COD/ (kg MLVSS · h)

An experimental determination of the specific substrate elimination rate needed to dimension the selector can only be used if the mixed liquor is adapted to the specific selector system and wastewater. These limitations only apply if an existing selector has to be optimized or when a pilot plant is operated with a selector. If investigations are done with a mixed liquor adapted to a selector, but not adapted to the specific wastewater the substrate elimination rate will very likely be underestimated and this will lead to an oversizing of the selector. However, if the selector is divided into several cascades it is probable that the selector will operate satysfactorily

The volume of a selector cascade can be calculated from equation 1 using an expected average load for S_S and a median specific substrate elimination rate that has been determined. If an experimental determination is not available, a median substrate elimination rate of 0.5 kg COD/(kg MLVSS·h) can be assumed. In these cases the oxygenation capacity must not be limiting and the concentration of S_S in the mixture of wastewater and RAS in the influent of the selector must be above 200 mg COD/L. If the concentration of S_S is below 200 mg/L, $q_{(SS,\ OUR)}$ must be reduced proportionally to the concentration of S_S, so that $q_{(SS,\ OUR)}$ is not lower than 0.2 kg COD/(kg MLVSS·h). The total selector volume is calculated from equation 1 with a maximum daily load of S_S and a minimum $q_{(SS,\ OUR)}$ in the first cascade. If no experimental measurements are available a value of 0.2 kg COD/(kg MLVSS·h) is recommended. When extreme peak loads of S_S are expected the selector volume must accommodate these peak loads. In such cases 50 % of the selector volume is calculated for the median substrate removal rate and the other 50 % of the selector is designed for the minimum substrate removal rate. In most cases the selector design under these conditions will have 2 to 4 cascades and a hydraulic retention time (based on influent and return sludge) on the order of 10 to 30 min. The oxygen demand in the selector can be calculated from equation 2:

$$OD = A_{COD} \cdot Q_{INF} \cdot S_{SINF} \qquad (\text{kg } O_2/\text{h}) \qquad (2)$$

Substrate removal rate, DO uptake rates can be determined during the experiment. In the absence of data on DO demand for substrate removal (A_{COD}), a value of 0.20 to 0.25 g O_2/(COD removed) should be used.

To prevent the storage capacity from being limiting in the selector, the minimum return sludge ratio should be:

$$\frac{Q_{INF} \cdot S_{SINF}}{Q_{RS} \cdot MLVSS_{RS}} \leq 0{,}2 gCOD / gMLVSS \qquad (3)$$

Q_{INF}	inflow (m^3/h)
$S_{S\ INF}$	influent soluble readily degradable substrate (g/m^3)
Q_{RS}	return sludge flow (m^3/h)
$MLVSS_{RS}$	Volatile suspended solid in return sludge (g/m^3)

Using the design value of 0,2 g COD/g MLVSS only ≤ 0,4 g COD/g MLVSS will be stored in the activated sludge.

At an F/M ratio > 0.5 kg COD/ (kg MLSS·d) or at a sludge age of less than 5 d the stored substrate will not be metabolized completely, and this will have an adverse effect on storage capacity and selector efficiency. For reliable prevention of bulking the COD loading rate should be less than 0.5 kg COD/(kg MLSS·d). A value of 0.3 to 0.4 kg COD/(kg MLSS·d) and a minimum sludge age of 5 d at 20°C are recommended. At lower temperatures a higher sludge age is recommended.

6.6.6.2 Storage capacity and substrate removal after shock loads, maximum storage capacity

After extreme shock loadings of the plant it has been observed that a selector at Leopoldsdorf plant removed only a small part of the substrate. The oxygen uptake and the removed COD load indicated that substrate was immediately metabolized and no storage took place. Based on the influent load and on the oxygenation capacity, removal and storage could be calculated. The storage amounted up to 0.3 COD/g MLVSS. An increase in the effluent COD concentration was not observed dung this event. The storage capacity of the sludge was nearly exhausted but not exceeded. Kroiss (1981) reported that in experiments with shock loads in the pilot plant at Leopoldsdorf up to 0.28 g COD/g MLVSS could be stored. Dohanyos *et al.* (1971) found that mixed liquor which was loaded only once per day with a shock load of glucose could store up to 0.4 g of polysaccharide per g of MLSS. In a mixture of peptone and glucose a maximum storage capacity of 0.3 g polysaccharide/g MLSS was observed.

6.6.6.3 Influence of SRT and F/M ratio on the substrate elimination in the selector

The activity of a mixed liquor expressed as maximum oxygen uptake rate is a function of SRT (solids retention time). With a reduction of the SRT the rate of storage and possibly the storage capacity of mixed liquor is increased. Because the storage capacity of the mixed liquor is limited when sludge age is reduced the substrate stored in the selector cannot be metabolized in the aeration tank, and in consequence this leads to the effect that no storage can take place in the selector. Investigations at the Leopoldsdorf treatment plant at an aeration tank temperature of 18°C showed that with a sludge age below 5 days or an F/M ratio above 0.3–0.4 kg COD/(kg MLSS·d) no substrate elimination by storage was observed in the selector. In the Hohenau treatment plant with aeration tank temperatures as high as 28 °C under comparable conditions a loss of the storage capacity in the selector was only observed at an F/M ratios above 0.5 kg COD/(kg MLSS·d) and sludge ages below 4 days. According to investigations of Dohanyos *et al.* (1971) the mixed liquor storage capacity reaches a maximum at a sludge age of 10 days. Sludge ages below 5 days showed a rapid decrease of the storage capacity. These results are in good agreement with the investigations of Eikelboom (1982) showing that above an F/M ratio of 0.5 kg BOD_5/(kg·d) biosorption significantly decreases.

6.6.6.4 Nutrient supply and substrate elimination

The storage of substrate in the selector under aerobic conditions is not influenced by the availability of nitrogen. However, for the metabolism of the stored material a sufficient supply of nitrogen and phosphorus is necessary. When nitrogen or phosphorus are deficient stored material cannot be metabolized, and the selector loses its efficiency.

6.6.6.5 Adaptation of mixed liquor to a selector

When an activated sludge plant is equipped with a new selector it takes about 1/2 sludge ages until the activated sludge develops significant storage capacity. The substrate removal rate by uptake for storage can exceed the steady state substrate removal rate at maximum growth rate by a factor of 3 to 10.

6.6.6.6 Microbial growth and storage of substrate in the selector

A COD balance on the selector and the entire treatment plant shows that only a small part of the COD removed in the selector is metabolized in the selector itself. The major part of this COD is stored. For example, a COD balance on the Hohenau treatment plant (Fig. 6.22) shows that only 15-20 % of the substrate removed in the selector is metabolized there (Prendl, 1997). An explanation for these results is given by Van Loosdrecht *et al.* (1997) where the storage of polymers like polyhydroxybutyrate and glycogen as internal storage material is described. Glycogen is important as storage compound when crucial for the type of bacteria present or in case carbohydrate is present in the wastewater. Those bacteria which can store the substrate with a high rate do not need to adapt their growth rate to the concentration or to the amount of substrate available at the moment but they are able to grow on the stored substrate with an average growth rate even in case low external substrate is available. These bacteria which are able to store substrate have an advantage in the development in case they are exposed to a cyclic change of surplus and deficit of substrate (e.g. plants with selectors or SBR-plants). They therefore will suppress those bacteria which are not able to store substrate. The storage can be performed under aerobic, anoxic and anaerobic conditions. The energy necessary for the storage of the dissolved COD can be obtained by the oxidation of a small part of the dissolved COD under aerobic or anoxic conditions. In an anaerobic selector the energy is produced by the hydrolysis of polyphosphate, by those bacteria that have this physiological property. The specific oxygen demand for the elimination of the dissolved easily degradable substrate in the selector is a function for the amount of energy which is necessary to convert the easily degradable substrate from the waste water into storage material in the cells.

In Austria, lab and pilot-scale experiments were initiated as early as 1972 to use selectors for treating sugar mill wastewater. Bulking sludge was cured by installation of a selector with a volume of 2.5% of the aeration tank and by nutrient addition. Removal of 70% of the influent COD load took place in the selector at a hydraulic retention time of 5 minutes. Additional 20% of the influent load was removed in the aeration tank: the overall COD-removal was > 90%. The oxygen consumption in the selector was 8 % of the COD removed, while 92% of the COD was stored in the sludge.

The conclusions drawn from these investigations were the basis for the design of a full-scale plant that started up in 1983. More than 10 years of operation verified the COD balance and the bulking sludge control. When bulking did occur (due to Type 021N) soluble readily biodegradable substrates (sugar and lactic, acetic, propionic, butyric acid) could be detected in the main aeration basin because not completely eliminated in the selector.

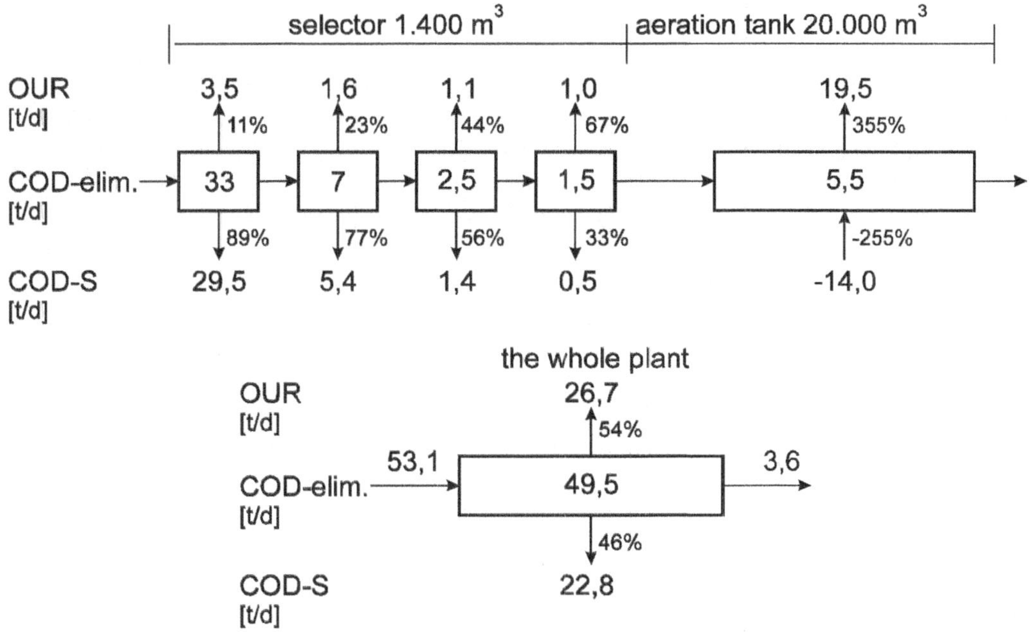

Fig: 6.22. COD-balance of Hohenau wastewater treatment plant, Nov. to Dec. 1995

6.6.7 Gram-positive filamentous bacteria in low load WWTP

Traditional microscopic investigations in Germany revealed a significant shift from Gram-negative filamentous and non-filamentous bacteria in the late eighties to Gram-positive bacteria prevailing now (Lind and Lemmer 1998, Müller *et al.* 2004, Lemmer *et al.* 2004). This group of organisms is frequently found at present in municipal low load nitrifying WWTP with long sludge age.

Excessive growth of these organisms accompanies sludge floating problems in the aeration tank, secondary clarifier, and also in digesters. This group of organisms includes in particular *M. parvicella* and nocardioform actinomycetes, and also morphotypes such as *Nostocoida limicola,* and Type 1851 (see Table 6.17). In certain nutrient conditions, (N/P and/or O_2 depletion, starvation conditions with shortage of carbonaceous substrate as in sludge stabilization plants) these organisms are known to form hydrophobic cell walls, that allows to readily adhere to and stabilize gas-water interfaces (Lemmer 1996). Investigations on both filamentous and non-filamentous bacteria isolated from scum, revealed many of these organisms to be able to produce biogenic surface-active material such as glycolipids. Thus they are able to promote both scum initialization and stabiliziation (Lemmer *et al.* 2000).

6.6.7.1 M. parvicella

From a taxonomic point of view it turned out that all *M. parvicella* strains investigated worldwide so far belong to one phylogenetic entity (Erhart *et al.* 1997), i.e. specific countermeasures being proposed should be effective worldwide.

In earlier investigations *M. parvicella* was shown to depend on long-chain fatty acids as a sole carbon source (Slijkhuis 1983) as well as on reduced sources of nitrogen, phosphorus, and sulfur. Troubleshooting measures aim at decreasing the input of grease and oils, synthetic surfactants, and other substrates providing long alkyl chains. In his work Slijkhuis (1983) showed that long-chain fatty acids, short-chain fatty acids also might enhance of *M. parvicella* growth yield. Thus we assume that our observations that digester supernatant promotes *M. parvicella* growth might not only be explained by the high ammonia concentrations, but also by the high input of short-chain fatty acids into the aeration tank (Lind and Lemmer 1998). *M. parvicella* was promoted, when particulate substrate was hydrolyzed in preceding anaerobic and/or anoxic zones. This makes long-chain fatty acids readily available for this organism (Eikelboom and Andreasen, 1995) and appears valid for the German WWTP surveyed.

In many cases increasing sludge load was an appropriate measure for decreasing *M. parvicella* growth. Knoop and Kunst (1998) showed in pilot and full scale plants *M. parvicella* was selected, when the F/M ratio was below 0.15 kg BOD_5/(kg MLSS·d) and the temperature was below 15°C. A temperature increase to 20°C made *M. parvicella* break apart into short fragments and become less abundant. Our recent work does not indicate the low temperature it self to be the selecting factor but rather the indirect decrease of both hydrolization and/or nitrification rates.

As *M. parvicella* is known to depend on reduced N-compounds (Slijkhuis 1983), ammonium shock loads could promote its growth and should be avoided. It was found to be beneficial to add digester supernatant in a balanced way to the aeration tanks. Moreover an NH_4 concentration < 1 mg/L at the end of the aeration period, as proposed by Eikelboom (2000) and Kruit *et al.* (2002), was successful to suppress excessive *M. parvicella* growth.

The oxygen affinity of *M. parvicella* is high so that the organism is competitive in N/DN and EBPR plants with oxygen-depleted conditions. Moreover, in situ determination of substrate uptake by microautoradiography has proved *M. parvicella* to be able to take up and store fatty acids under anaerobic conditions (Andreasen and Nielsen 2000), a feature which makes *M. parvicella* strongly competitive in German EBPR plants.

M. parvicella was sensitive to aluminum salts. In many cases phosphate precipitation with Al-salts was preferable to Fe-salts at concentration of 2-3 mg Al/(g MLSS d) as proposed by Eikelboom and Andreasen (1995) and Lind and Lemmer (1998): this was effective in controlling *M. parvicella* (Paris *et al.* 2004). In general sour salts such as $AlCl_3$ proved to work better than basic salts such as Na-aluminate. The real mechanism of Al-dosage seems to be an inhibition of lipases being situated outside *M. parvicella* cell wall. This was demonstrated by enzyme labelled fluorescence (ELF): after incubation with a fluorescently labelled long-chain fatty acid substrate, a fluorescent precipitate shows up at the lipase activity site. Al-dosage investigations with *M. parvicella* dominated sludges showed a marked decrease of fluorescent precipitates, i.e. lipase activity decrease (Schade and Lemmer 2004).

6.6.7.2 Nocardioform actinomycetes

Nocardioform actinomycetes in previous investigations belonged to various genera. Besides different temperature preferences, however, they seem to be quite similar in physiological properties. This means that countermeasures against the "morphotype" might be a sufficient means for practitioners.

In the mid 1990s we observed by classical microscopical methods that in Germany branched filamentous actinomycetes did not any longer play the important role as they did about 10 years before (Lind and Lemmer 1998). We thought this might have been due to the low load conditions in present nitrifying WWTP. Certainly they need a long mean cell residence time to not be washed out the system. However, as was shown in previous investigations, actinomycetes require much higher substrate concentrations for excessive growth than is necessary for many other Gram-positive bacteria from low load WWTP (Lemmer 1996). Meanwhile we confirmed by gene probes that nocardioform actinomycetes are still well established in scum biocenoses. However, they do not necessarily grow as filaments but occur as bent rods, as single cells or in cell clusters (Müller *et al.* 2004). Moreover despite their common nickname "Nocardia" no such genus was found in activated sludge, while the genera *Rhodococcus* and *Dietzia* were commonly found in German WWTP (Müller 2005).

In contrast to other Eikelboom morphotypes comprising both taxonomically and physiologically very different bacteria, nocardioform actinomycetes seem to be controllable by similar pragmatic troubleshooting measures despite their occurrence as diverse genera. Empirical observations show that actinomycetes occur with high loads of hydrophobic substrate such as grease and oil. Thus decreasing such substances load is a stable and efficient countermeasure. Moreover, because nocardioform actinomycetes happily grow in high sludge age scum layers (Müller *et al.* 2004), it is crucial to continuously remove floated sludge from tank surfaces.

6.6.8 Eikelboom morphotypes *Nostocoida limicola*, Types 0041/0675, 1851, and 0092

Scum bacterial components in German WWTP comprise various Eikelboom morphotypes besides *M. parvicella* and nocardioform actinomycetes (Lemmer *et al.* 2004). These morphotypes, initially described by Eikelboom (1975) on the base of morphological criteria and Gram and Neisser reaction, belong to various bacterial genera and species, and even clusters as is known today. We determined the phylogenetic affiliations of this group in our WWTPs by molecular probe techniques and confirmed their diversity. This diversity makes at present impossible to recommend specific troubleshooting measures for these "non-famous scum bacteria". Their diversity also explains the scores of inconsistent practitioners' reports on the countermeasures tested successfully. Present microbiological knowledge about scum bacteria others than *M. parvicella* and nocardioform actinomycetes forces us therefore to practice pragmatic troubleshooting (Lemmer *et al.* 2004).

Empirical observations show that *N. limicola* morphotypes preferentially proliferate at moderate load conditions (approx 0.2 kg BOD_5/kg MLSS·d). Large amounts of this organism are encountered in systems with anaerobic zones and with DO deficient conditions. If nutrient removal systems have to remove high sugar loads, *Actinobacteria* species of *N. limicola*, morphotype proliferate (Schade and Lemmer 2002). These problems could be controlled by a specific treatment.

Selection factors of Types 1851 and 0092 are still not clearly defined: some physiological characteristics and troubleshooting measures have been summarized (Lemmer and Lind, 2000). When only oxidized N-sources are available, Type 0092 might outcompete *M. parvicella,* Type 1851, or *N. limicola* morphotype *Actinobacteria* strains, that occur during full nitrification in summer time. Type 0092, as a Gram-negative organism, makes its contribution to scum stabilization by cell hydrophobicity much less serious and easier to handle. This is also true for Type 0041/0675 which proliferate at low loading conditions (0.01 to 0.1 kg BOD_5/kg MLSS·d) and seems to prefer particulate food over soluble substrate. Thus the organism is competitive in systems without premixing raw wastewater zones and recycling sludge, i.e. in systems where substrate hydrolyzation in the primary effluent occurs at low rates. Here we can confirm literature data (Eikelboom and Andreasen 1995).

As a pragmatic but nonetheless helpful measure we advise our practitioners to work on a shift of scum populations from effective floaters such as *M. parvicella* and nocardioform actinomycetes to ineffective scum stabilizers such as the morphotypes described above. Their scum is much easier to handle as it can be sprayed down and does not cause severe problems in anaerobic digesters as it is for *M. parvicella* and actinomycetes (Lemmer *et al.* 2004).

6.6.9 Non-specific control measures for sludge bulking and scum

Unspecific control measures for sludge separation problems do not aim at inhibiting or eliminating specific filamentous organisms. They help without knowing exactly the biogenic causes. This includes sludge treatment with chemicals in order to weigh down bulking sludge or damage filaments as compared to flocs, or calculated sludge handling such as removal of floating sludge fractions. They are known for long and partially described in the report of the ATV-DVWK task group KA 6.1 (Lemmer *et al.* 1998).

6.6.9.1 Addition of aluminum salts for phosphate precipitation

Simultaneous precipitation with Al salts in N/DN Plants with plug flow regimes in many cases is a successful control measure, that allows to reduce the amount of filaments in the sludge and the SVI (Eikelboom and Andreasen 1995). Actinomycetes, *M. parvicella and Sphaerotilus* spp were sensitive to Al salts (see chapt. 6.6.7.1).

6.6.9.2 Addition of layered silicates

Layered silicates can weigh down activated sludge (Matsché and Kreuzinger 1998). The talc powder addition has been recommended. It is incorporated into sludge flocs and does not separate in either the aeration tank or the secondary clarifier. In a Type 021N bulking case with prior SVI values of 800 mL/g the addition of talc (at a dosage of talc/biomass about 0.6:1) stabilized the SVI at 100 – 150 mL/g after 14 d (Eikelboom and Grovenstein 1998). However, at a high dosage (> 50 % of talc compared to sludge) or in WWTP with low sludge ages the high talc dosages increase the amount of surplus sludge to be disposed (Kroiss *et al.* 1997). Nonetheless talc has been applied as an efficient countermeasure against bulking events in some German plants.

6.6.9.3 Addition of lime

Lime addition is recommended as an emergency control measure. Slaked lime or lime hydrate can be dosed into the RAS so that the sludge is exposed to a pH value of 10.5 to 11 for a short time (Matsché 1991). Filamentous bacteria experience more severe damage than single cells which are protected in the sludge floc matrix. Treatment efficiency recovers after about 2 d and the loss for this period of time is less deleterious than long-term sludge losses from bulking.

6.6.10 Decreasing the biomass of filamentous bacteria

Filamentous sludge bacteria that accumulate in floating sludge such as *M. parvicella* and nocardioform actinomycetes can be removed by thorough removal of the floating sludge. Effective systems are available now for both scum removal from aeration tanks and sludge layers removal from secondary clarifier surfaces. Scum and floating sludge should not be pumped into the digester because similar foaming problems may occur there (Kunst and Knoop 1996, Lind and Lemmer 1998). Sludge centrifugation and disposal on drying-beds has been successful. Investigations are developing to obtain floating sludge reduction strategies such as the thermophilic disintegration. Recent experiments showed that the application of specific detergents (Bioversal) improved significantly the scum settleability, allowing to transform floating scum into settleable sludge, which can be further treated in combination with the excess sludge (Matsché 2003).

6.7 ITALY (C. LEVANTESI, V. TANDOI)

6.7.1 General situation

There are approximately 8500 municipal wastewater activated sludge plants in Italy and other 3000 under construction (Istat 1996); there are also a large number of industrial wastewater activated sludge plants.

The plants have mainly aerobic or anoxic/aerobic configuration. At present very few EBPR plants exist but the introduction of this kind of plant may increase depending on the definition of "sensitive areas", where more restrictive EU limits for phosphorus and nitrogen might apply. 30 to 40 % of activated sludge plants in Italy suffer from bulking and foaming problems.

Microscopic observation of activated sludge is now a common practice, especially in larger plants, and some data are available on the main filamentous bacteria present. These are identified using the manuals of Eikelboom and van Buijsen 1983, Jenkins et al. 1993 or Italian versions of these manuals (Cingolani and Ciccarelli 1996, Madoni et al. 1996)

6.7.2 Filament surveys

Several surveys have been made on the filaments responsible for bulking and foaming problems, in domestic plants in Southern Italy (Di Marino et al. 1992), Central Italy (Rossetti et al. 1994), in industrial plants (Molinari and Canziani, 1992). Recently two wider survey were made all the country around (Madoni et al. 2000) and in Northern Italy WWTP (Pasinetti and Emondi, 2005).

Table 6.18. Filament and chimico phisical parameters in bulking plants (Di Marino et al., 1992)

Plant	Persons equival.	pH	Redox P.,MV	BOD mg/L	COD mg/L	TKN mg/L	P mg/L	SVI mL/g	Filaments in abundance order
Andria	84000	6.51	-250	2730	7000	102	30	298	*Thiothrix*>Type 021N>Type 1701>*Nocardia*
Bari Ovest	371000	7.94	-298	630	1220	87	10	326	*Thiothrix*>Type 021N>*Nocardia*
Bari Est		7.47	-196	488	1041	85	14	332	Type 021N>*Thiothrix*>Type 1701>*Nocardia*>*Type 0041*
Brindisi	62000	7.96	-301	360	760	45	12	307	*M. Parvicella*>*Nocardia*>Type 021N>Type 0092>*Thiothrix*
Casalnuovo	2800	8.37	76	307	700	71	9	521	Type 021N>Type 0675>Type 0041>*M. Parvicella*
Cassanom.	7800	7.42	-55	350	805	75	28	523	Type 021N>Type 1701>*Thiothrix*>Type 0041
Castelnuovo	2300	8.1	-122	430	1280	54	14	530	*Thiothrix*>Type 021N>Type 0041>*M. Parvicella*
Cerignola	51000	7.28	-58	645	1510	89	23	341	*Thiothrix*>Type 1801>*H.Hydrossis*>*Nocardia*
Chieuti	1900	7.12	-198	490	890	57	11	612	Type 021N>Type 0041>Type 0092>*M. Parvicella*
Crispiano	11500	7.46	-147	430	888	52	15	403	Type 021N>*Thiothrix*>H. *Hydrossis*>*S. Natans*
Gioia Colle	27000	6.41	-54	1380	7300	227	62	349	*Thiothrix*>Type 021N>Type 1701>*N. Limicola*
Ginosa	21700	8.31	-50	470	880	108	13	510	*Thiothrix*>*H.*Hydrossis>Type 0961>Type 1701 *Nocardia*
Giovinazzo	21000	7.58	-61	555	1180	111	15	588	*Thiothrix*>Type 1701>Type 021N > H. *Hydrossis*>Type 1851
Gravina	36000	8.12	68	594	1926	74	10	300	Type 021N >*H.*Hydrossis>*N. Limicola*>Type1701
Grumo	12000	7.77	-82	585	1057	62	19	411	Type 1701>*H. Hydrossis*>Type 0092>Type 0041
Minervino M.	12200	8.47	-80	486	917	50	1.5	505	Type 021*N.*>*Thiothrix*>*N .Limicola*>*Nocardia*
Molfetta	64400	7.98	-290	552	1117	171	22	459	*Thiothrix*>*M. Parvicella*>Type 1701>Type 1863
Novoli	9000	7.2	-180	280	410	62	12	350	*M. Parvicella*>*Thiothrix*>Type 1701
Ascoli S.	7200	7.0	-150	460	830	56	10	290	S. Natans>Type 1701>*Thiothrix*>Type 1863>*H. Hydrossis*
Volturara A.	1100	7.4	-50	320	560	42	7	500	*H. Hydrossis*>*N. Limicola*>Type 1701>*Nocardia*>Type 1851
Villa Castelli	7500	7.31	-193	444	742	52	14	367	*M. Parvicella*>Type1701>S. Natans>*N. Limicola*
Monteparano	2100	7.85	10	519	1718	82	7	343	*M. Parvicella*>*Thiothrix*>Type 0041> Type 0092>Type 1863
Sammichele	7000	6.38	-138	498	1089	84	9	305	*Nocardia*>*Thiothrix*>Type 021N Type 0041>*N. Limicola*
Nola	27000	7.38	-245	492	1028	85	7	285	*Nocardia*>*Thiothrix*>Type 021N>Type 1701>Type 1863

6.7.2.1 Puglia Region, Southern Italy

Approximately 30 % of the 172 plants examined had problems, and severe filamentous bulking (SVI > 300 mL/g, filament abundance 5 or 6) was the most common problem encountered (24 cases).

The main microorganisms causing problems (Table 6.18,) were *Thiothrix* (dominant in 9 and secondary in 10 cases), Type 021N (dominant in 7 and secondary in 8 cases), *M. parvicella* (dominant in 4 and secondary in 4 cases), and *Nocardia* (dominant in 2 and secondary in 8 cases). *Sphaerotilus natans* and Type 1701 were dominant only in one case and others (*Haliscomenobacter hydrossis*, Type 0041, *Nostocoida limicola* and Type 1863) were present only as secondary filaments. The sewage very often had a low redox potential which may favour the abundance of *Thiothrix* and Type 021N.

Foaming was a problem in 5 cases (caused by *M parvicella* and *Nocardia*) and mainly resulted from long sludge age and recirculation of scum to the head the plant.

6.7.2.2 Lazio Region, Central Italy

This survey considered 39 activated sludge plants located in the Lazio Region of central Italy (Rossetti et al. 1994). Problems with the activated sludge process were extensive. Of the plants surveyed, 46 % had filaments at abundance levels *very common*, 30 % *abundant* and 5 % *excessive*.

The main filamentous bacteria present in the plants were *Microthrix* and Nocardioforms (dominant in 32 and 15 % of the plants respectively) and Type 0092, Type 0041 and Type 0675. The watewater characteristic was quite different from Southern Italy, probably because the wastes were not as septic.

6.7.2.3 Industrial wastes, Northern Italy

Five plants with more severe problems were selected from amongst 15 N/DN plants treating wastes from various industries (fine chemicals, tannery, petrochemical, pharmaceutical, food processing). The main filaments were Type 0041 (dominant twice, and secondary three times), *Nostocoida limicola* (dominant once, secondary three times), Type 0092 (dominant once, secondary twice), *Haliscomenobacter hydrossis* (dominant once), and a group of other organisms found only as secondary filaments (Type 0675, *Nocardia* spp., *M. parvicella*, Type 1701, Type 1851, Type 1863). These data again confirm that the organisms that causes the main problems of bulking and foaming, *M. parvicella* and *Nocardia*, play a minor role in activated sludge fed with industrial wastewater (Molinari and Canziani, 1992).

A very recent survey (Pasinetti and Emondi, 2005) has been done on 86 WWTP, mainly industrial (70), in a wide F/M range(0.2-0.8 kg COD kg/kg SS d). The situation in the country improved from the past years: only foaming is a problem in 36% of domestic WWTP while no bulking problems have been recorded in the others. Bulking is a problem only in 21 % of industrial WWTP (no foaming problem and never GALO or *M. parvicella* are detected recorded in such kind of WWTP). In domestic plants the more abundant organisms were GALO, Type 0041/0675, *M. parvicella*, *N. limicola*, *Thiothrix*, Type 0092. A wide variety of filaments was found in industrial WWTP: the more abundant organisms were Type 0041/0675, *N. limicola*, *Thiothrix*, Type 0092, plus other such as *S. natans*, Type 1701, Type 0803, Type 1851, Type 1863, Type 0914, H. hydrossis.

6.7.2.4 Survey on 167 Domestic plants.

Five hundred plants had been contacted in Italy, receiving 167 answers, complete of data and sludge samples (Madoni et al. 2000). 66% of plants were affected by bulking or foaming, and 33% suffered both problems. *M. parvicella*, Type 0041 and Type 021N were the most common filaments in bulking sludges, while foam were mostly dominated by *M. parvicella* and GALOs. *M. parvicella* and Type 0092 dominated either conventional and BNR plants, while *Nostocoida limicola* and Type 1701 dominated only in conventional plants. The main problems remarked by plant operators was the deterioration of effluent quality, due to the effluent suspended solids increase.

6.7.3 Control methods

Both specific and non-specific methods for controlling bulking and foaming phenomena have been utilised in Italy.

6.7.3.1 Non specific methods

Coagulants: Coagulants such as $FeCl_3$ (50 mg/L), Ca $(OH)_2$ (50 mg/L), are used, especially in small treatment plants, and often in conjunction with cationic polyelectrolytes (1-2 mg/L, Vismara 1998).

Toxic compounds: Sodium hypochlorite is used for bulking control at a dosage of 5-15 mg/L in the RAS. Usually the sludge is exposed at least three times per day to the effect of the chlorine (Cingolani and Ramadori 1992, Beccari et al. 1995). The use of H_2O_2 at doses of 10-15 mg/L (Vismara 1998), is very limited because it is not as readily available as sodium hypochlorite.

6.7.3.2 Specific methods

Specific methods of bulking control are usually based on the correlation between filaments and cause (Jenkins et al. 1993). Success was also obtained by using "Contact zone" (2 solved cases on 2 tested), nutrient addition (7 solved cases on 8 tentatives), antifoaming addition (6 solved cases on 7 tentatives), coagulant addition (24 solved cases on 33 tentatives), sludge age decrease (40 solved cases on 75 tentatives), dissolved oxygen decrease (19 solved cases on 30 tentatives), mechanical removal of foam (17 solved cases on 27 tentatives), etc. (Madoni et al. 2000).

6.7.4 Case studies at full scale plants

6.7.4.1 Reduction of Type 021N by chlorination at the Bagnuolo (Reggio Emilia) municipal plant

The plant in Bagnuolo receives sewage from 9,000 PE (90 % domestic, 10 % industrial). The plant has a secondary clarifier built inside the aeration basin, and the RAS is by an airlift.

The plant suffered severe bulking problems (DSVI > 500 mL/g) with a secondary clarifier sludge blanket normally at 10 cm. Filaments were present at abundance level 5-6, with Type 021N dominant and Type 1701, Type 0041 and *M. parvicella* secondary. $FeCl_3$ had been added (unsuccessfully) so chlorination was tested as control strategy (Davoli and Guglielmi 1998).

The remedial action

Sodium hypochlorite was added at rate of 3 g Cl_2/g SS d directly into the aeration basin. The DSVI decreased after a week to 300 mL/g. The abundance of the filament was constant for a long time (class 5-6) and decreased only after 45 days (class 4). The filaments were damaged by the chlorine addition showing empty spaces trough the trichomes. Damage first appeared on Type 1701, then Type 021N and *M. parvicella* latterly.

Effect of chlorine on respiration rate

Oxygen uptake rates were not affected by chlorine addition. The heterotrophic activity (measured adding acetate and NH_4 nitrogen) showed values of 20-30 mg O_2 / g SS h, while the autotrophic activity (due only to the oxidation of NH_4 nitrogen) was 3- 8 mg NH_4-N / g SS h throughout the period of experimentation.

6.7.4.2 Controlling growth of Nocardia and M. parvicella by selectors at the Moncasale (Reggio Emilia) treatment plant

The use of a selector for controlling the growth of *Nocardia* and *M. parvicella* was tested in the full-scale plant of Moncasale (Reggio Emilia). The plant designed for 280,000 EP (domestic and small amount of industrial waste) is composed of three lines of 2,000 m^3 each. The plant suffered severe foaming problems due to *Nocardia* (summer) and *M. parvicella* (winter). Three experiments were carried out to evaluate the use of a selector (Davoli et al. 1997).

Effect of anoxic selectors on Nocardia growth.

An anoxic selector of 250 m^3 was installed in the first part of the first line: a second line was used as control.

The selector produced a temporary decrease of *Nocardia* content, but eventually an increase of *M. parvicella*. Two main conclusions were drawn:

- The selector organic load (1.8 kg BOD_5 kg^{-1}MLSS d^{-1}) was not high enough to favour the growth of floc formers, over the filaments
- *M. parvicella* tended to increase as *Nocardia* decreased. It appeared that the growth of this organism was due to the disappearance of *Nocardia*, which is normally present at the higher temperatures.

A smaller anoxic selector (100 m^3) was created to obtain a higher control organic load.

The 100 m^3 anoxic selector effectively controlled *Nocardia* and it was eliminated from the plant in 40 days. *Nocardia* remained in the control plant (at level of 2.5 10^6 intersections/μl). The selector reduced all filamentous bacteria level and the DSVI was 50 – 120 mL/g compared to 80 – 300 mL/g in the control plant.

Effect of a 100 m^3 aerobic selector on the growth of M. parvicella.

Because it is known that *M. parvicella* can grow in plants with unaerated zones, an aerobic selector was used. The aerobic selector controlled but did not eliminate it (30 – 40 filaments intersection/ μl in comparison with 40 – 80 in the control plant. The DSVI ranged from 100 – 300 mL/g compared to 250 – 400 mL/g in the control plant. Also the foam on the control plant settling tank surface was higher than in the control plant.

6.7.4.3 Experiences at Varese – Olona plant (northern Italy) for controlling Nocardia and M. parvicella growth by reducing sludge age

The treatment plant of Varese-Olona has experienced significant foaming problems on the aeration basin and final sedimentation tank: SVI values were typically 250 mL/g and microscopic observations showed very small flocs dominated by *Nocardia* and *M. parvicella.*

The strategy adopted (Bargna et al. 1993) was to reduce the sludge age from 15 to 2 days, over a two months period. A progressive improvement in both the quality of the effluent and the settling characteristics of the sludge occurred. The SVI stabilised under 200 mL/g and COD removal efficiency increased. The flocs gradually became more compact and *Nocardia* and *M. parvicella* eventually disappeared. Nitrification capacity was lost but the nitrogen removal by biomass synthesis was enough to produce an effluent that complied with Italian limit for N (15 mg /L).

6.7.4.4 Type 021N and Thiothrix reduction by an anoxic selector at Area Nolana Treatment Plant

Configuration and operating parameter of the treatment plant.

The Area Nolana Treatment Plant is designed for 450,000 P.E. and currently receives 1/3 its design flow. Several rectangular aeration basins constitute the biological section. Sludge age is normally 4 days and the organic load is 0.3 kg BOD_5 kg^{-1}MLSS d^{-1}. Like many plants in Southern Italy the RAS is introduced in the primary settling tanks, and the waste sludge is taken from the bottom of the primary settler (a mixture of primary and secondary sludge). This procedure causes two problems:

- it is very difficult to control activated sludge. The sludge age is normally higher that expected
- fermentation occurs in the primary tank

The effluent of the primary tank normally has a very low ORP (-125 mV) and contains high concentrations of short chain fatty acids and sulphides, that can promote the growth of the filaments Type 021N and *Thiothrix.*

Significant problems occur in the summer when the seasonal industries (tomato canning) produce a large increase of VFAs in the sewage. In this period the settling properties of the activated sludge deteriorate dramatically (DSVI up to 900 mL/g) and the content of filaments Type 021N and *Thiothrix*, increases significantly.

The solutions adopted

Chemical addition ($FeCl_3$ 40 g/m^3 for 10 d) and sodium hypochlorite addition (3 g Cl_2/kg SS d to the RAS for 15 d) did not produce any settling improvement.

An anoxic selector (250 m^3, 1/24 of the volume of oxidation basins) was installed (Cesaro 1997) utilising a basin to transfer the mixed liquor, by three screw pumps to the aeration basin. The recirculation flow and the influent mixed together in this selector. This simple solution was implemented by using existing facilities.

The selector hydraulic retention time (influent plus recirculation flow) was 7.4 minutes, and the NO_3-N in the recirculation flow 5 – 7 mg/L. At the same time the sludge age was reduced (to 2.5 days) and the aeration basin DO increased (3.5 mg/L and 5.8 mg/L at the head and the effluent end.

Results obtained
The modifications introduced were kept for 80 days (industrial waste overload). The DSVI (initial value 800 mL/g) decreased after 15 d and after 35 d reached 190 mL/g. The ORP in the selector was approximately - 110 mV, indicating anoxic conditions.

75 % of the influent RBCOD (161 mg/L) was taken up in the anoxic selector. The selector mainly affected Type 021N populations which decreased from abundance level 6 to 1–2. The effect was less pronounced on the *Thiothrix* population that decreased from an abundance level 6 to 3–4. It is well known that Type 021N effects more the settling properties, because of the bridge effect on the flocs.

6.7.4.5 Selector investigations on four large WWTP in Central Italy

Full scale application have been tested in four large WWTP in Central Italy, suffering seasonal bulking and foaming problems: the effect of anoxic selectors has been investigated (Davoli et al. 2002). Plant A, C and D operated in plug flow conditions, while plant B as CSTR. The selectors worked with a soluble COD concentration in the range 14-69 mg/L, and were effective in two cases Plant A and C, where there was the lower amount of soluble COD measured in the selector effluent.

6.7.4.6 M. parvicella elimination by poly aluminium chloride at Fusina treatment plant

The Fusina (VN) biological treatment plant treats wastewater from the industrial and civil area of Venice, for approx. 330.000 equivalent inhabitants. During the past winters the plant suffered intense biological scum formation, due to the presence of the bacterium *M. parvicella* (Miana et al., 2002).

The plant was carefully monitored in the years 2001-2002, in order to individuate the reasons for the *M. parvicella* proliferation, and to find solutions for containing its proliferation. A positive correlation was found, once again, between the temperature and the presence of the organism in activated sludge (Fig. 6.23), and consequently the percentage of aeration basins covered by scum, in the winter period. No correlation was found with either surfactant and oil content in the influent; their concentrations in the influent did not change substantially during the course of the year.

The solutions adopted. In February 2002, a specific compound was tested, polyaluminium chloride, whose efficacy has been pointed out (Eikelboom, 1997).

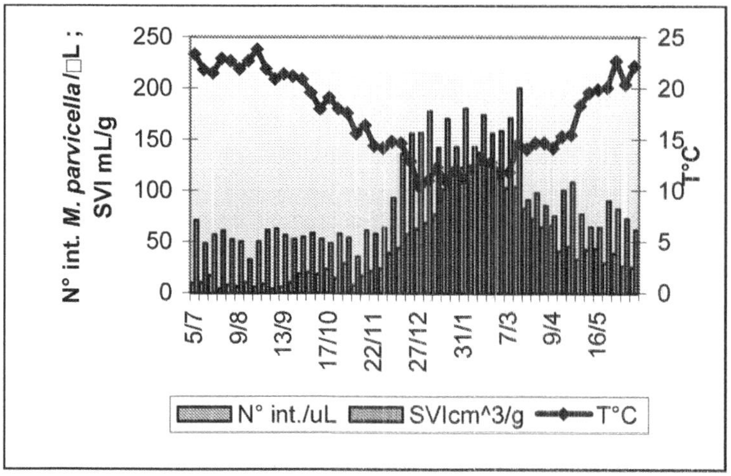

Fig. 6.23. *M. parvicella*, SVI and temperature trend during the period of experimentation 2000-2001 (Badoer et al. 2003).

Results obtained

The dosage of 1,5 g Al/ kg MLSS d in return sludge, for a period of 42 days, caused, already after two weeks dosage, a dramatic reduction either of the number of free filaments present in the mixed liquor and the foam covered surface (Fig. 6.24). The SVI decreased from 188 to 88 mL/g and analogously the suspended solids in the final effluent (from 60 to 24 mg/L) (Badoer et al. 2003). The *M. parvicella* filaments during the treatment appeared damaged and eventually were incorporated in the biological flocs. Considering that *M. parvicella* reaches high levels in the sludge only for a limited periods of time, normally in winter, the strategy of monitoring its presence, intervening with the polyaluminium chloride when necessary, appears effective, considering the reasonable cost and the limited days of the coagulant (or toxic) dosage.

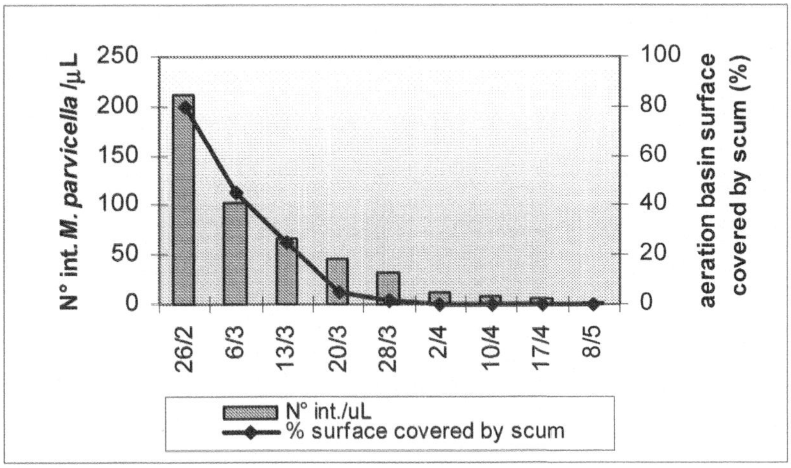

Fig. 6.24. *M. parvicella* presence, and aeration basins percentage covered by foam, during the polyaluminium chloride addition, February/May 2002 (Badoer et al. 2003).

6.8 JAPAN (T. MIMO)

6.8.1 General situation

There are 1,187 municipal wastewater treatment plants operating in Japan, out of which 93 percent are the activated sludge plants (as of 1995, Sewage Database of Japan). Many of these have bulking or foaming problems. Japan Sewage Works Agency (JSWA), which is a semi-public agency supervising municipalities on design and operation of sewage systems, conducted three comprehensive surveys of operational problems in Japanese activated sludge processes; the first survey was in 1982-1984 (JSWA, 1983; 1984; 1985), the second in 1993 (JSWA, 1994) and the third in 1997-1999 (Yumikura and Ishi, 1998). Key observations from these surveys are summarized in Table 6.19. Type 021N was always the No.1 filament causing bulking, and *Nocardia* spp. was the No.1 organism causing foaming. Type 021N bulking had little relation to operational conditions such as BOD-SS loading, temperature or DO, and tended to occur in winter or during periods of seasonal temperature change (either spring or autumn). Seeding of Type 021N and *Nocardia* spp. from sewers into WWTP contributed significantly to their proliferation. Septic sludge return liquor and septic primary effluent promoted Type 021N growth.

Table 6.19 Summary of Comprehensive Surveys on Activated Sludge Separation Problems in Japan.

Year of Survey	1997	1993	1984	1983	1982
Method of Survey	Survey on Sewage Database of Japan containing info from 824 WWTPs, and questionnaire to selected 200 WWTPs suffering from bulking (181responded).	Questionnaire to 785 WWTPs for general survey (663 responded) and to 405 suffering WWTPs for detailed survey (318 responded).	Investigation into type of filaments in aeration tanks and sewage pipes, operational conditions and influent quality in 54 WWTPs.	Identification of filaments which caused bulking in 8 WWTPs and measurement of *Nocardia* in inflowing sewage at 5 WWTPs.	Questionnaire to 45 WWTPs (42 responded), identification of foam/scum causing fila-ments in 22 WWTPs, and further detailed study in 9 WWTPs.
Major Observations	- Each WWTP often has its own seasonal pattern of occurrence of bulking; for example, high occurrence in snow melting season or in the period of temperature change (spring or autumn). - Eikelboom Type 021N is the most dominant, but in many cases there are additional sub-dominant species including *Sphaerotilus*, Type 0041, *Thiothrix*, Type 0675, etc. - Introduction of the anaerobic-aerobic process could effectively control Type 021N bulking.	- 82 % WWTPs experienced separation problem, and 32 % suffered from filamentous bulking. - Possible causes of bulking include: temperature change in spring or autumn, inflow of septic backwater from excess sludge treatment line, winter season, high loading, low DO. - 82% of all the bulking were caused by Type 021N. - Possible causes of foaming/scum formation include: long SRT and high MLSS, excess aeration, summer season, circulation of scum within the plant, - Major foam causing species was *Nocardia amarae*.	- The most dominant filaments in WWTP suffering from bulking were Type 021N, followed by Type 1701, *Nocardia amarae*. - BOD-SS loading, temperature, DO or geological location did not have significant relation to proliferation of Type 021N. - Growth of Type 021N in sewage pipes, leading to supply of seed of 021N to WWTPs, as well as supply of sulfide from primary sedimentation tank or excess sludge treatment lines, seemed to promote Type021N bulking.	- Type 021N mainly caused bulking. - High *Nocardia* number ($2.4 \cdot 10^3$ cells/mL) detected in sewage to a foaming plant, cf. less than $1 \cdot 10^2$ cells/mL at non-foaming plants.	- 52% suffered from bulking. - 74% suffered from foaming, out of which 59% caused by actinomycetes and 14% due to denitrification or septic conditions. - *Nocardia amarae* appeared at temperature range of 14-26 ℃, Whereas *Rodococcus* spp. appeared around 26 ℃ in warm area or in summer. - Nocardia seemed to be inoculated from sewer pipes. - Higher SRT (Av. 9.1 d) provided higher risk of *Nocardia* foaming than lower SRT (Av. 5.2 d).

Sources: 1997 survey - Yumikura and Ishi (1998); 1993 survey – ISWA (1994); 1984 survey – JSWA (1985); 1983 survey – JSWA (1984); 1982 survey – JSWA (1983)

6.8.2 Major Filaments Responsible for problems

The major filaments appearing in activated sludge plants in Japan are summarized in Table 6.20. Results of a survey of filamentous microorganisms in Thailand (Mino, 1995) are also given in Table 6.20.

Table 6.20. Frequently observed filaments causing bulking in Japan and Thailand

Country	Japan								Thailand
Survey	JSWA, 1997		JSWA, 1993		JSWA, 1984		JSWA, 1983		Mino, 1990
No. of WWTPs	195		162		43		11		6
Filaments	Rank	Appearance %	Rank	Appearance %	Rank	Appearance %	Rank	Appearance %	Presence
Type 021N	1	41	1	81.5	1	42	1	73	++
Nocardia sp.	NA	NA	2	45.0	2	28	2	36	+
S. natans	2	11	3	11.5	5	9			
Type 0041	3	5	4	6.3	6	7	3	9	++
Thiothrix sp.	4	3	8	0.7					
Type 0675	5	3							
M. parvicella					4	14			
N. limicola					6	7			+
Type 1851	6	2	6	4.5					
Type 0961	7	2	5	5.2					
Type 1702	7	2					3	9	
Type 1701	9	1	7	3.4	3	19			++
H. hydrossis					6	7	3	9	+
Type 0092									+
Not Identified	-	29			-	30			++

NA: Not available. Sources: 1997 survey - Yumikura and Ishi (1998); 1993 survey – ISWA (1994); 1984 survey – JSWA (1985); 1983 survey – JSWA (1984); Mino 1990 survey – Mino (1995).

6.8.3 Research on Fundamental Aspects of Filamentous Bulking in Japan/Thailand

Kohno (1987) carried out a bacteriological study of filamentous microorganisms in activated sludge from municipal and industrial WWTPs. He collected activated sludge samples, and tried to identify and classify observed filaments. According to his classification, there are five major groups of filaments (Kohno Type I – V), each of which is further divided into several sub-groups. Kohno (1987) described each sub-group morphology and physiology in detail. Kohno Type II-3 can be identified as Eikelboom Type 021N using Eikelboom's identification key (Eikelboom and van Buijsen, 1983). However, Kohno (1988) indicated that the observed morphology of Japanese Type II-3 was not exactly the same as Type 021N described by Eikelboom. A filament in Thailand was definitively identified as Type 021N based on its morphology (Mino, 1995). Although the cell shape of the Type 021N was square and barrel like the usual 021N, the irregular curvature typically seen for Type 021N were hardly seen. The morphological differences observed between Type 021N from different geological areas imply that there is significant diversity in the so-called Type 021N. Kanagawa et al (1998) investigated the 16S rDNA of the five strains of Type 021N isolated by Kohno, and found that they were phylogenetically diverse and that the sequence of the oligonucleotide probe for Type 021N developed by Wagner et al (1994) and the corresponding sequences in 4 strains of Kohno's isolates had 1 or 2 mis-matches.

6.8.4 Control Measures Applied

Through the JSWA surveys (1983, 1984, 1985, 1994), the causes of Type 021N bulking were identified as follows:

1) Seeding of Type 021N cells from sewers into activated sludge plants.
2) Septic conditions in sewers which lead to production of short chain fatty acids (SCFAs) by fermentation and generation of sulfides by sulfate reduction.
3) Carbon-rich industrial wastewater.
4) Septic conditions in the primary sedimentation tank
5) Septic return flows from sludge treatment process.

Possible counter measures for the Type 021N bulking are:

1) Proper operation of pump stations in the sewer network or proper control of influent gates to WWTP: accumulation of organic sediments in the sewers should be avoided because it may create septic conditions. Flushing sewers by proper operation of pumps or influent gates can prevent their septic conditions.
2) Bypass primary sedimentation tanks: too long a retention time in primary settlers may lead to septic conditions at the bottom of the tanks. Bypass of the primary settler sometimes improves bulking conditions.
3) Introduction of anaerobic-aerobic or anoxic-aerobic reactor configurations: since Type 021N is an obligate aerobe and does not denitrify, contact of activated sludge with substrates under either anaerobic or anoxic conditions can suppress it. Both anaerobic-aerobic processes for EBPR and anoxic-aerobic process for N/DN can provide such conditions. Introduction of anaerobic or anoxic zones to conventional activated sludge has proved effective for Type 021N bulking control. Even reduction of air supply to the influent end of the aeration tank is sometimes effective.
4) Physico-chemical measures: both RAS line chlorination and addition of coagulants to aeration tanks are effective methods for controlling filamentous bulking. When chlorine or coagulants addition is stopped, the activated sludge bulking can return.

6.8.5 Case Studies

The case study (JSWA, 1986) describes how the bulking was cured in a conventional activated sludge plant (capacity of 16,591 m^3/d and average influent BOD 70-90 mg/L). The history of bulking at the plant is summarized in Table 6.21. The responsible filament was not identified but, judging from the general characteristics of the phenomenon and the SVI values observed, Type 021N was probably the causative organisms.

During the period 1976 – 1982, several measures were undertaken. Addition of digested sludge or local soil was not very effective. Development of new activated sludge biomass in an aeration tank was attempted, but was not successful. In 1983, the primary sedimentation tank was bypassed and unsettled wastewater was introduced directly into the aeration tank. This improved the bulking situation a little. In the mean time a coagulant was added and this led to improvement, but it costs about 3 million yen for a three month period. In 1985, a combination of several measures was tried including wastewater pumping strategy and introduction of anaerobic zone at the influent end of aeration tanks. To avoid accumulation of sludge in the influent pipes, the water level in them was maintained as low as possible. In addition, aeration in half of the aeration zone was significantly reduced to minimize oxygen supply. After these changes were applied, the bulking conditions were consistently improved and SVI was maintained below 200 mL/g.

Table 6.21 History of bulking at Plant T (JSWA, 1986).

Year	Average SVI (mL/g)	Highest SVI (mL/g)	Duration of Bulking Period (days)	Bulking Period
1976	113	343	29	Dec – Jan (77)
1977	108	730	22	Dec – Jan (78)
1978	99	187	0	-
1979	136	482	66	Dec – Mar (80)
1980	113	316	21	Nov
1981	239	649	185	Sept – Mar (82)
1982	171	654	66	May – June
1983	219	852	114	Nov – Feb (84)
1984	179	698	46	Sept - Nov

Acknowledgement

The author would like to express his sincere thanks to Mr. Yumikura Junichi of Japan Sewage Works Agency (JSWA) for providing comprehensive information obtained through surveys by JSWA.

6.9 SOUTH AFRICA (A.R. PITMAN)

6.9.1 General situations

The activated sludge process was introduced into South Africa in 1921/2. Most of the plants were conventional activated sludge facilities treating settled sewage. The first bulking problems were noted at the Johannesburg Bruma Works in 1932/3. These were attributed to low DO in the aeration tank. This problem was eliminated when re-aeration of the return sludge was introduced (Osborn, 1988). Most of the activated sludge separation problems with these first-generation plants were due to either low DO bulking sludge or to sludge floatation in final clarifiers due to unintended denitrification. These aeration problems and other operational difficulties, led to the activated sludge process falling into disfavour in the country. Nitrifying biofilter plants were then built to achieve the 10 mg/L ammonia-N limit which became a legal requirement in the early 1960's.

6.9.1.1 Extended aeration plants

The late 1960s and early 1970s saw a return to the activated sludge process. To comply with the new 10 mg/L ammonia-N standard, these second-generation plants were designed to nitrify with many using the extended aeration process with long sludge ages up to 30 d. Most of the smaller ones used the carousel configuration. During the operation of these plants, aerators were turned off during periods of low organic loading to save power and achieve denitrification within the reactor instead of the clarifiers. By creating suitable anoxic and anaerobic conditions at the inlet of these reactors, BNR was obtained. A major drawback of this procedure was the occurrence of severe bulking and foaming/scum problems, particularly in the late winter/early autumn months. In fact, bulking and scum problems came to the fore in South Africa with the advent of biological nutrient removal (BNR).

Typical examples of these were the Johannesburg' Alexandra and Olifantsvlei extended aeration plants with reactors in the form of a number of sequential aerobic compartments, divided by surface cut-off baffles. In both of these plants, aerators were switched off at the inlet of the reactor to create anaerobic/anoxic zones for biological nutrient removal (Nicholls, 1975, Venter, *et al.*, 1978). Severe foaming and scum formation was experienced at the Alexandra plant, particularly in the winter months. *Nocardia sp.* was the predominant filamentous organism in both the scum and the bulk mixed liquor. The next most dominant was Type 0041. Late winter DSVI levels ranged between 120 and 180 mL/g. In contrast at the Olifantsvlei plant winter DSVI never went above 100 mL/g and only mild foaming problems occurred. The predominant filaments were Types 1851 and 0041. However, when Olifantsvlei was modified to treat settled sewage and the feed and return sludge were premixed before entering the reactor, scum formation in winter became much more severe and DSVI increased into the range 150-180 mL/g. *Nocardia sp.* was then the predominant filament with *M. parvicella* and Type 0092 the secondary filaments (Pitman, 1996).

Similar problems were experienced at the Umhlatuzana plant in Pinetown. This had a single compartment extended aeration reactor in which aerators were switched off at the inlet and at night to create anaerobic/anoxic conditions. Severe bulking was experienced with DSVI of up to 300 mL/g. Typical experience with the carousel configuration is at Oudtshoorn, where bulking occurs year round. SVI are in the range 130 to 400 mL/g and scum formation is a problem. Predominant filaments are *M. parvicella, Nocardia* spp., and Type 1701 (Little, 1998). Similarly, Bux and Kasan (1994) found *M. parvicella* and Type 0041 to be the most prevalent filamentous organisms in a carousel plant in Durban.

6.9.1.2 Conventional activated sludge

Most of the conventional activated sludge plants in South Africa are single aeration basin nitrifying plants treating settled sewage at relatively long sludge ages. Typical of the bulking problems encountered is the experience at the Durban Southern Works. Here sludge SVI values range between 230 and 250 mL/g, which coincide with the proliferation of *Nocardia* spp. foam/scum (Hilditch, 1998).

Severe bulking problems were also experienced at Fishwater Flats in Port Elizabeth. Here settled sewage was treated in a single compartment aerobic reactor operated at a long sludge age and liquid retention time 24h. Predominant filamentous organisms responsible were Types 021N and 0041 with 1851 and 0092 being commonly ranked second/third. When aerobic selectors with a retention period of 0.5 h were retrofitted at the reactor inlet, the problem was dramatically reduced. Although bulking still occurs, DSVIs rarely exceed 200 mL/g with the typical range being 100-150, with peak values occurring in winter. Moderate foaming has also occurred due to *Nocardia sp.* and to a lesser extent *M. parvicella* (Wakefield 1998).

6.9.1.3 *Plants treating industrial wastes*

There are few activated sludge plants treating purely industrial effluents in South Africa and even fewer have been reported on in the literature. Van Leeuwen (1989) reported an interesting case of zoogleal bulking in a wastewater treatment plant at a liquid fuel synthesis complex. This fuel from coal plant produced an effluent with a very high soluble COD content.

6.9.2 Plants specifically designed for biological nutrient removal (BNR)

6.9.2.1 *First generation plants*

The first generation plants specifically designed for BNR were either the 4-stage Bardenpho or 5-stage Phoredox/Bardenpho configuration. Unfortunately, bulking and foaming problems continued with these plants. Typical experience was with Johannesburg's 5-stage, Goudkoppies (GK) and Northern Works Unit 3 (NWU3) plants: both plants experienced prolific growth of *M. parvicella* scum in the late winter/autumn months. At NWU3, Type 0803 filaments dominated in the mixed liquor in early winter with *M. parvicella* the second most dominant. As the winter progressed, the situation was reversed with *M. parvicella* becoming the most dominant. This coincided with an increase in bulking and scum formation. At Goudkoppies, early winter saw Type 0041 followed by Type 0092 as dominant filaments.

Again as winter progressed, *M. parvicella* became dominant as the bulking and scum formation intensified. The control of scum growth on these older BNR plants was very difficult as the baffles separating the various zones inadvertently trapped scum, causing the surface layer to have a far longer retention period than the bulk mixed liquor. In addition during the early operating phase of the Johannesburg plants scum trapping was exacerbated by the usually fruitless exercise of creating additional anoxic zones at the start of the main aerobic zone in an attempt to improve denitrification. Under these circumstances, DO levels in the aerobic zone were depressed and quasi-aerobic conditions were created. This caused severe bulking problems with DSVI values in excess of 250 mL/g and very thick (>300 mm), dried-out scum layers on the unaerated zones (Pitman 1996).

A noticeable exception amongst the earlier BNR plants is the Johannesburg Bushkoppie plant (commissioned in 1982). During the whole of its operating life, this plant has never experienced any bulking or foaming problems (Pitman, 1996). DSVI values have always been below 100 mL/g with the average value for October (one of the worst months of the year), being 60 mL/g. Higher DSVI values, (up to 90 mL/g) are experienced whenever the incoming peak-load exceeds the aeration capacity available (i.e. low DOs). Important features of this plant are an open surface flow through the reactor and an anaerobic zone split into two sequential compartments each representing about 5% of the total reactor volume.

6.9.3 Filamentous organisms in South African activated sludge plants

In a survey of 111 activated sludge plants, Blackbeard et al. (1986) found that 56 plants produced bulking sludge (DSVI >150 mL/g). Of all the plants surveyed, 26 were nutrient (N&P) removing and the other N removing. The filamentous organisms causing the bulking problems on these plants were generally of the so-called low F/M type. In a follow-up survey, Blackbeard et al. (1988) concentrating on 33 N and P removal (BNR) plants, found that 27 produced bulking sludge. The six most frequently dominant filaments in decreasing order of dominance, found in both studies, were Types 0092, 0675, 0041, *M. parvicella,* and Types 0914 and 1851. In investigations directed specifically at the foam on these plants, they found in both studies that the most frequently dominant filaments were Type 0092, *M. parvicella* and *Nocardia* spp.

In a different study of 10 activated sludge plants in Natal, Bux and Kasan (1994), found that the most frequently occurring filaments were *M. parvicella* (100% of plants), *Type* 0041 (60%), and *Nocardia* (50%).

Detailed studies in Johannesburg (Pitman, 1996) have shown that the most common scum producing bacteria on BNR plants is *M. parvicella*. In addition, when scum infections reach their peak in the spring months, *M. parvicella* is also the most common filamentous organism in the bulk mixed liquor as well.

6.9.4 Reasons for the growth of filamentous organisms in SA plants.

The reasons that have been put forward for sludge bulking in activated sludge plants in South Africa are the usual ones such as low DO, low F/M, poor reactor configuration etc.

In their survey of bulking and foaming in activated sludge plants in South Africa, Blackbeard et al. (1986) noted that foam is often trapped in bio-reactors and clarifiers because of inadequate foam removal facilities. In addition foaming is usually exacerbated by the physical retention of foam in the plant. If foam is continuously removed from the plant and not inadvertently returned, foaming problems could be significantly reduced.

Surface trapping by baffles which hinder surface flow through the reactor has been shown to be the major cause of the proliferation of scum forming filamentous organisms in the process (Pitman, 1996). In this regard the multiple reactor compartmentalization needed in the BNR process can greatly exacerbate this problem. In many of the older BNR plants the baffles dividing the reactor into the various anaerobic, anoxic and aerobic zones were such that they stood well above the top water level in the reactor closing off the complete surface to surface flow through the reactor. Floating material was then trapped for prolonged periods in the various zones, particularly the unaerated zones. The scum retention period was often well in excess of the reactor sludge age, allowing the scum forming organisms to proliferate.

Research at the University of Cape Town has put forward another possible reason for excessive filamentous growth in BNR plants. Six years of laboratory scale pilot studies focused on filamentous bulking in BNR plants (Casey *et al.*, 1992a, Casey *et al.*, 1992b, Casey *et al.*, 1994, Casey *et al.*, 1995, Ekama, *et al.*,1996a, Ekama *et al.*, 1996b, Gabb *et al.*, 1996a, Gabb *et al.*, 1996b, Still *et al.*, 1996). A comprehensive evaluation of the selector effect (induced either through the use of selector reactors or via intermittent feeding) and its influence on low F/M filaments, showed that selectors (whether anaerobic, anoxic or aerobic), did not control the growth of many of the low F/M filaments found in South African full scale activated sludge plants. Thus they could make no conclusion on the role of selectors in the control of this type of filaments (Lakay,1999).

Their experiments showed that low F/M filament proliferation was ameliorated by continuous aerobic and continuous anoxic conditions, but was promoted by intermittent aeration or alternating anoxic-aerobic conditions.

Based on laboratory-scale research and a literature review of the biochemical pathways of aerobic-facultative heterotrophs, they proposed a hypothesis for the proliferation of anoxic-aerobic (AA) filamentous organisms in BNR plants as follows (Casey *et al.*, 1994, Casey *et al.*, 1999):

"In activated sludge, under anoxic conditions, floc forming organisms carry out the denitrification of nitrate (NO_3^-) through each of the denitrification intermediates to dinitrogen (N_2), in the process of which the intermediate nitric oxide (NO) is accumulated intracellulary. Intracellular NO is inhibitory to the utilization of oxygen in the subsequent aerobic zone. In contrast filamentous organisms carry out only part of the denitrification pathway, i.e. the reduction of NO_3^- to NO_2^-, they do not accumulate NO and hence are not inhibited in the subsequent aerobic zone. Thus in anoxic-aerobic systems, floc-formers are placed at a disadvantage in the aerobic zone giving an advantage to the filaments in the competition for substrate."

A biochemical/microbiological model was developed to illustrate this hypothesis which was supported by experimental evidence. They also proposed a tentative strategy for the control of AA filamentous bulking, which was tested successfully at laboratory scale. In this they proposed that a BNR process be operated such that the intracellular NO was reduced to near zero at the end of the anoxic zone. In doing this the floc formers would not be inhibited in the downstream aerobic zone. As NO could not serve as a control parameter they proposed that NO_2^- be used as a surrogate parameter as NO precursor in the denitrification pathway. To prevent high concentrations of NO_2^- at the end of the anoxic zone the plant should be designed and operated to achieve complete denitrification in the anoxic zone (Casey *et al.*, 1994).

Although the hypotheses of the UCT Group have been proved fairly conclusively at laboratory scale there is a need to carry out full-scale studies into this phenomenon.

6.9.5 Methods used for the control of bulking and foaming in South Africa

6.9.5.1 Physical removal

Many operators have tried manual removal of scum and clarifier water-sprays. Once removed, treating the removed material directly in the sludge handling plant prevents re-inoculation. These methods have proved to be very tedious and of limited value (Johannes 1998). This is because the design of the activated sludge reactor is such that the growth of scum producing bacteria is encouraged.

Selective flotation as a mechanical means of scum removal was examined by Pretorius and Laubscher (1987) using a pilot-scale diffused-air flotation cell. They found that over 95% of the scum forming organisms could be removed from activated sludge plagued by *Nocardia* sp. within 4 hours. Although their study showed that this technique could remove and control the overabundant occurrence of scum-forming organisms in activated sludge, they nevertheless emphasized that every attempt should be made to minimize in-plant environmental conditions which would favour these organisms growth.

6.9.5.2 Use of toxicants

Chlorination

Chlorination is a commonly used non-specific bulking control method. Lakay et al. (1988) studied the control of bulking in a pilot-scale modified UCT BNR plant. They found that this technique was satisfactory provided the guidelines of Jenkins et al. (1984) were followed. Although they found that chlorination produced a relatively minor loss in nutrient removal efficiency, dosing had to be carefully controlled. Of the filamentous organisms in the plant, *M. parvicella* showed the greatest resistance to chlorination, with doses up to 4 mg Cl_2/g SS d, having minimal effect.

The first successful use of chlorine for full-scale bulking control was at the Daspoort sewage works in Pretoria (Thirion 1982). In follow-up studies on this BNR plant, Saayman et al. (1996,1997) showed that 1.42 g Cl_2/kg MLSS d reduced the DSVI from about 180 to 105 within 10 days; but their target DSVI <100 could not be reached. However, when the dose was increased to 3.07 g Cl_2/kg MLSS d, the DSVI reduced to 80 mL/g in 14 days. At the lower chlorine dose there was minimal detrimental effect on biological nutrient removal, however at the higher dose bio-P removal was inhibited.

Successful chlorination has also been carried out in other activated sludge plants (Cooper, 1998, Johannes, 1998, Borain, 1998 and St Arnaud, 1998), although it is recognized as a temporary solution.

Ozonation

The use of ozone for bulking control has been extensively investigated at both pilot and full-scale. In studies on a pilot-scale BNR plant van Leeuwen (1988) showed that an ozone dose of 2 g O_3/kg MLSS/d reduced the DSVI from 180 to less than 100 mL/g, within one sludge age, with no disturbance to the BNR process. Even doses of up to 30g O_3/kg MLSS.d produced no decrease in N and P removal. van Leeuwen and Pretorius (1988) found that filamentous growth in a pilot-plant could be suppressed by a continuous dose of 4 g O_3/kgMLSS.d into the aerobic zone. At this dosage ozonation was not significantly more expensive than chlorination. In limited full-scale studies, Saayman and van Leeuwen (1992) found that an ozone dose of 0.3 to 0.6 g O_3/kg MLSSd reduced the average DSVI from 180 to 120 mL/g. At this dose a temporary inhibition of phosphorus removal was observed but no significant inhibition of nitrogen removal.

Hydrogen Peroxide

Limited research work using hydrogen peroxide for bulking control was carried out at the University of Pretoria. In bench scale studies, van Leeuwen (1990) found that a hydrogen peroxide dose of 8 g H_2O_2/kg MLSS d was required for bulking control. A continuous dose between 5 and 10 g H_2O_2/kg MLSS d into the aerobic zone had no adverse affect on the BNR process. In fact massive doses up to 100 g H_2O_2/kg MLSS·d did not appear to be detrimental to the BNR biomass.

6.9.5.3 Reduction of sludge age

Reducing the sludge age is a commonly used procedure to try and solve an activated sludge separation problem. It has the dual effect of reducing the flux load on clarifiers and suppressing the growth of low F/M filaments (Lakay *et al.* 1999). This procedure has been successfully utilised on many plants in the country, such as Hilditch, 1998, Cooper, 1998, King, 1998. This option is sometimes not available in the winter months due to the need to operate at higher sludge ages to guarantee nitrification.

6.9.5.4 Elimination and prevention of surface trapping

As mentioned above, surface trapping is the main case of the growth of scum producing filamentous organisms in activated sludge reactor. Great success has been achieved in eliminating the nuisance growth of scumming/bulking filamentous organisms by designing the reactor so that no trapping of the surface layer occurs (Pitman 1996, 1997). Some older plants, which inadvertently cause scum trapping, are being modified to eliminate this problem; for example the Johannesburg Northern Works Unit 3 and the Oudtshoorn carousel reactor (Little, 1998). The alterations to the Northern Works Unit 3 reactors were very successful. Surface trapping of scums was eliminated and the average DSVI values decreased from over 200 to less than 130 mL/g.

6.9.5.5 Increase in clarification capacity

As some plants in the country have been designed in anticipation of operating with a good settling sludge, the construction of additional clarifiers is often the only way of dealing with a sludge bulking problem (King, 1998).

6.9.5.6 Adjustments of dissolved oxygen levels

Low DO bulking has been experienced on plants where the incoming load has exceeded the aeration capacity in the reactor. This occurs when the load to the plant has exceeded its aeration capacity or when aerators are deliberately switched off to create anoxic zones. In addition low DO conditions can occur in the inlet of a 'plug-flow' aerobic zone. This usually occurs under peak load conditions. A good example of this occurred at the Johannesburg Bushkoppie plant, where diffuser blockages reduced the ability of the plant to handle peak loads. The problem was overcome by installing new diffuser domes and increasing the diffuser density at the inlet of the aerobic zone.

6.9.5.7 Improvements of reactor configuration

Much of the earlier investigations into the BNR process involved increasing the unaerated volume fraction of the activated sludge reactor as a method of improving N and P removal. However subsequent experience showed that this could be detrimental to sludge settleability. Because of this Pitman (1991) proposed that the unaerated mass fraction of the BNR reactor be limited to a maximum 0.4. In addition Ekama *et al.* (1996b) showed that low F/M filaments will proliferate in full-scale plants with unaerated fractions in excess of 0.5.

Casey *et al.* (1999) suggested that a small aerobic or anoxic reactor receiving a fraction of the feed, be placed immediately upstream of the main aerobic zone to ensure the removal of intracellular NO which had been shown inhibit the growth of floc forming organisms in the aerobic zone (Casey *et al.*, 1999).

One of the main reasons for bulking and scumming problems in activated sludge plants is a reactor configuration that inadvertently encourages the growth of filamentous organisms (Pitman, 1996). Full-scale research in Johannesburg has shown that it is possible to design the BNR reactor such that the growth of filaments is suppressed (Pitman, 1991, 1994, 1996). Details of this are described later in this chapter.

6.9.6 A case study

The experiences with bulking and foaming problem on the Johannesburg works provide an interesting case study as they are mirrored by similar problems elsewhere in the country.

All the activated sludge plants in Johannesburg are designed and/or operated for biological nutrient removal with supplementary iron chloride addition. Basically two types of bulking problem have occurred on these plants:

Low DO bulking which has occurred when DO concentrations in all, or portions of the aerobic zone were too low. Low DO bulking is quite infrequent and usually easy to cure.

Scum formation with related bulking sludge has been a far more common problem which has not always been easily solved. This was due to the fact that most of the earlier designs of BNR reactors inadvertently encouraged the growth of scum forming filamentous organisms.

Experiences with the latter are as follows:

- Bulking sludge and foam/scum formation usually occur at roughly the same period of the year.
- The occurrence of bulking and foaming follows a seasonal cycle with the first signs of the problem starting in autumn (May), a build-up through the winter to reach peak severity in early spring (August). During the late spring the problem diminishes with minimal problems being experienced in summer (October-March).
- Heavy summer rains tend to signal the final breakup and disappearance of the scum.
- The predominant scum producing organism is *M. parvicella,* which is also the predominant filamentous organism in the bulk of the mixed liquor.
- Scum formation always occurs in activated sludge reactors, which have zone dividing baffles, or scum baffles that restrict surface flow through the reactor from inlet to outlet. This results in trapped surface layers which have a retention period far in excess of the liquid, or even the solids retention time in the reactor. Two solids phases then occur, the surface layer/s and the bulk of the mixed liquor passing submerged apertures in the baffles.
- On unaerated (anaerobic and anoxic) zones with surface trapping dividing baffles, scum accumulates profusely, to produce thick mats covering the complete surface These tend to dry out and cause odour problems.
- Another mechanism for scum retention in unaerated zones is the surface flow patterns of mixers. This can also cause scum to be trapped around mixer shafts for prolonged periods. This occurs (although to a lesser extent), even where there is no flow trapping baffle at the end of such zones. The greater the mixer power input, the greater the magnitude of this effect.
- Scum also accumulates in areas of the aerobic zone where surface aerators are switched off to save power. This does not occur in diffused air systems.
- Bulking tends to lag scum formation. For example, the first appearance of scum on the reactor does not coincide with any significant DSVI increase. However when the scum formation reaches significant proportions, (i.e. covering 50% of the reactor surface) the

DSVI tends to increase. Conversely at the end of the winter/autumn season the scum often has to virtually disappear before the DSVI values will drop back to low levels (DSVI less than 70 mL/g). This appears to indicate the contamination of the bulk mixed liquor from the scum layer.

- Experience in Johannesburg has shown that it is possible to design activated sludge reactors that do not encourage undesidered scum growths and bulking sludge. Such reactors have the following features:
- A semi-plug-flow configuration with highly compartmentalized anaerobic and anoxic zones.
- A free surface flow through the reactor from the inlet to the outlet. This means that there is no surface trapping and the surface layer has much the same retention period as the bulk of the mixed liquor. This is achieved by ensuring that the division walls between the various zones have surface gaps or are fully submerged below the surface. In addition the design of the aeration system, mixers and points of flow withdrawal and entry into zones are such that scum is not trapped by hydrodynamic effects or recycled within the reactor, via mixed liquor recycles.
- Scum removed from secondary clarifiers is not recycled to the inlet of the works but is fed to the waste activated sludge thickening plant.
- To date ten rectors, each with a treatment capacity of 50 Ml/d (ADWF) have been built using these features. These have operated for many years with minimal or no scum formation and stable DSVI values which never exceed 100 mL/g and are usually in the range of 50-70 mL/g.

6.10 UNITED KINGDOM (R. FOOT AND M. ROBINSON)

6.10.1 General situation

Early experiments with activated sludge waste-water treatment in the UK (Ardern & Lockett, 1914) have since been developed, researched and applied throughout the world. Much of this work is documented in scientific and engineering journals. The problem faced by design engineers and operators has been to select from this wealth of information, design data and process control to give value design and security of treatment. In the UK the Activated Sludge Handbook of UK Wastewater Practice (CIWEM Publications, 1997) is used for standard design. However control strategies for bulking and scumming appear less specific but often follow the principles set out in Tomlinson and Chambers' (1984) benchmark paper "Control strategies for bulking sludge".

An understanding of the groups of filamentous organisms responsible for bulking and scumming (Wanner & Grau, 1989) has led to control specific to their metabolic preferences, while benefits from identifying the main species phylogenetically (Blackall *et.al.* 1996) have yet to be realised. At the same time the definition of the physico-chemical properties of foam forming sludges (Kocianova *et.al.*, 1992) has gone some way to explaining how scum forms and control strategies work.

6.10.2 Current situation

In the absence of a national survey or recent published information it could be concluded that sludge bulking and foaming is no longer a critical issue in the UK. Investment coupled with current best practice for control has contributed to the present situation. Having said that it would be unreasonable to assume that the problem has gone away. More a case of it being managed in such a way as to not adversely affect effluent quality or plant operation (Foot & Robinson, 2003).

Tipping (1995) concluded in his operators overview of the activated sludge process that "The operational and ecological factors which are responsible for inducing microbial growth, sufficient for foam formation, are largely unresolved and likely to be site specific." Which is why physical control measures are frequently used with the expectation that the problem is likely to recur. While the temptation exists to "bend the biology" of activated sludge by expecting more from the process than it can deliver, what we now have is a greater awareness of the requirements (and hence limitations) of the organisms involved in waste treatment.

For example, paper waste treatment is a major industrial process and Type 021N a bulking filament commonly present. Deficiency of nutrients or trace elements in some industrial waste treatment processes (Gostic, 1995; Godley, 1998) increases the risk of bulking. Although it is possible to redress the nutrient balance in such treatment plants, the 'quick fix' dose with a toxicant is often used to control bulking.

An informal approach to most UK water companies revealed very similar findings to those experienced in the Wessex Water region. Activated sludge bulking and scumming is invariably associated with:

- industrial effluents or domestic waste with a large proportion of industrial waste
- plant loading
- plant configuration
- process control

As elsewhere in the world what has not changed are the types of filamentous organisms responsible for bulking and scumming, the nocardioforms and *M. parvicella* (Pitman, 1996).

Achieving full compliance with the European UWWTD (Urban Waste Water Treatment Directive) is likely to introduce additional pressures on activated sludge based treatment systems. Anticipated problems associated with biological phosphorus removal have not materialised with a general adoption of chemical phosphorous removal in the UK. The next challenge is to provide full treatment on restricted sites in coastal areas. Compact processes such as batch reactors, aerated filters and membrane bio-reactors will be required to treat sewages with increased chloride levels (due to saline intrusion). Increased septicity will also occur due to new treatment plant being located some distance from the population centre and the presence of sulphate from the saline intrusion. Unless controlled, these factors will increase the tendency to activated sludge bulking at such coastal sites.

6.10.3 Control measures

When confronted with a bulking or scumming event general practice in the UK water industry is to:

- review process and controls and make required corrections

- employ short term remedial action if necessary.

If the problem persists:

- review plant design
- implement design changes

Most problems have been resolved by adopting good operational practice (dissolved oxygen control, plant loading, correct recycle rate and MLSS concentration). New treatment plant include initial contact zones (selectors) to discourage filamentous growth, and problem sites are modified to include 'selectors' (Foot *et.al.*, 1993, Foot & Forster, 1997). A comparison of design parameters used by two UK water companies is shown in Table 6.22

Table 6.22. Oxic selector specifications

Design Parameter	value
Example A	
Retention time (minutes) - FFT	10
- DWF	15
Floc loading (g/kg)	80
Example B	
Retention time (minutes) - average sewage flow	15
Loading (kg/kg.d) - Total	3
- Initial zone	12
Velocity (m/s)	0.2
Aeration (kg O_2/m^3 d)	4
Dissolved oxygen (mg/L)	0.1-1.0
Status	oxic

Scumming on aeration lanes is not perceived as a problem unless it carries forward to settlement tanks or smothers surface aerators. Some companies use baffles to retain scum and/or spray bars on the final settlement tanks to return the scum to the mixed liquor. Scumming has been known on some industrial waste activated sludge plants and has been controlled by spraying final effluent directly onto the aeration tank. Chemical control is generally limited to industrial waste treatment.

Sufficient oxygen supply to treat peak daily load is essential for sludge stability thus making an allowance for aerator failure. This is critical in oxidation ditches where a single aerator failure represents a significant loss of process air. One site which receives a significant proportion of dairy and laundry waste treats effluent to a 5 mg/L ammonia standard with a residual DO of less than 1 mg O_2/L. High loading, a selective substrate and a long solids retention time ensures a background population of *Nocardia* sp and *M. parvicella.* Aerator failure results in immediate scumming. While loading can be reduced by chemically aided primary settlement at a cost of 30 pence per additional kg BOD removed the most viable option proved to be the provision of additional aeration at a revenue cost of 4 pence per kg/BOD removed.

In contrast, the following case history is interesting because the waste-water treatment plant was new and well designed for sludge control. However, when a bulking event occurred following an instrumentation fault, the normal control option (reducing mixed liquor solids) was not available.

6.10.4 Case history

6.10.4.1 Background

Holdenhurst waste-water treatment works is located in southern England and treats mainly domestic sewage from a population of approximately 170,000 inhabitants. The works is divided into two treatment streams, stream 1 is a contact stabilisation plant described previously (Foot *et.al.* 1993) and stream 2, a conventional activated sludge plant. The second stream was commissioned in 1994 and takes one third of the incoming sewage flow. The outline specifications of the treatment works are summarised in Table 6.23.

Table 6.23. Treatment works specifications for stream 2

Design Parameter	Value
Average flow (m^3/d)	14500
Average BOD (mg/L)	216
Total selector volume (m^3)	414
Initial zone volume (m^3)	82
Loading (kg/kg.d) - Total	13.2
- Initial zone	2.6
Aeration tank capacity (m^3)	8900
Organic loading rate (kg/kg d)	0.12
Aerators	diffused air (fine)
Aerator power (kW)	1080
Consent BOD (mg/L)	15
Consent N-NH_3 (mg/L)	5

The 'selector' was economically incorporated into the main plant as part of the flow distribution system. One common initial contact zone feeds two secondary zones which distribute flows to the six aeration lanes (Figure 6.25). Submerged mixers were installed to prevent settlement and the build-up of scum.

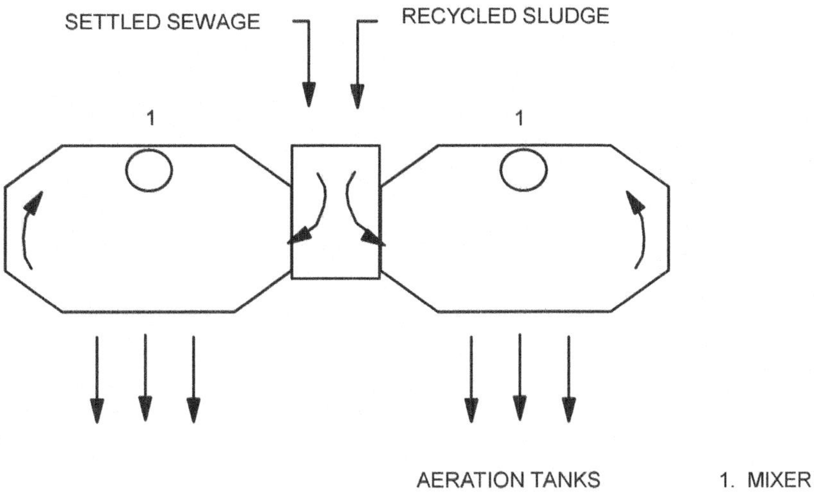

Fig. 6.25. Schematic diagram of the selector at Holdenhurst waste-water treatment works. Arrows indicate direction of flow.

Normal operation of the treatment stream ensures a SVI of less than 110 mL/g from an average mixed liquor concentration of 2.6 g/L. Figure 6.26 shows that from September to December 1996 the monthly average SSVI increased from 80 mL/g to 120 mL/g (maximum SSVI = 160 mL/g). *M. parvicella* was the dominant filament in the sludge (Eikelboom category 4) bridging sludge flocs, and increasing sludge bulk. In contrast there were very few filaments in stream 1 and the maximum SSVI was 110 mL/g. The main concerns were:

- elevated sludge blanket levels in the final settlement tanks
- reduced thickening performance at the sludge treatment plant

Effluent compliance was not an issue. With a sludge age greater than twenty days and sufficient settlement capacity neither nitrification nor solids settlement were at risk. However there was a drop-off in thickening performance in the sludge treatment works.

It is the plant operators job to control mixed liquor solids by maintaining sludge settlement within prescribed limits on an stirred jar settlement test. The surplussing rate is increased or decreased to keep within a percentage settlement band on the measuring cylinder but when it becomes impossible to keep within these limits without excessive surplussing, the process requires investigation. The subsequent site log for this event is described below.

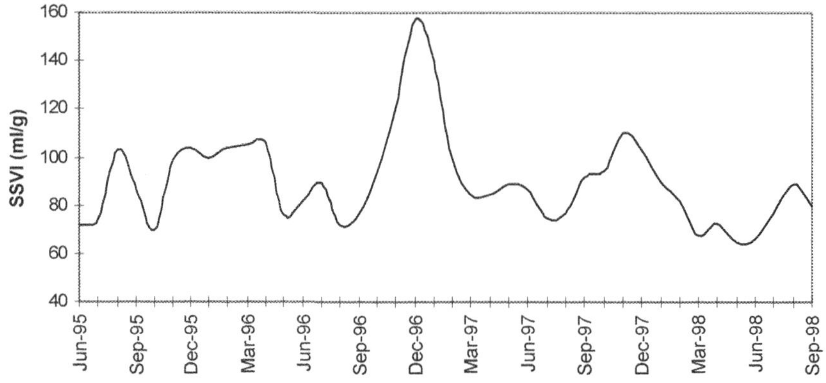

Fig. 6.26. Monthly average mixed liquor SSVI values for Holdenhurst waste-water treatment works (stream 2).

6.10.4.2 Solution Process

The process review confirmed plant loadings, identified plant faults and detailed the remedial action required to resolve the problem.

- faulty dissolved oxygen controllers
- F/M = 0.08 kg/kg.d when MLSS = 2.7 g/L and SRT = 30 days
- available process air = 0.56 - 1.68 kg O_2/m^3 d
- air required to avoid foaming (Foot *et.al.* 1994) at 0.08 kg/kg d > 1.2 kg O_2/m^3 d, at 0.12 kg/kg d > 1.6 kg O_2/m^3 d

This indicated over-aeration and low F/M bulking suggesting the following remedial action:

- repair faulty DO controllers
- increase F/M to 0.12 kg/kg d by reducing aeration volume to 7417 m^3 and then 5933 m^3 by taking lanes out of service
- maintain MLSS at 2.5 g/L and SRT 18 days by wasting solids at a rate of 2.7 L/s.
- maintain aeration in lanes out of service

The results of these actions are illustrated in Figure 6.26. Average monthly SSVI was reduced to 110 mL/g within two months during which time normal plant operation was resumed. Filament concentrations reduced to similar levels to the unaffected stream 2 in both the operational and non-operational lanes. No further capital expenditure was required and average monthly SSVI values have remained below 120 mL/g.

6.10.5 Conclusion

It is generally felt that bulking and scumming in the UK has been contained but not eliminated. Events are often associated with extremes of plant loading and process control. Industrial waste treatment or plant treating a large proportion of industrial waste present the greatest problems.

Considering the fact that the activated sludge process has to cope with varying loads and conditions and that compliance with effluent standards is required to be constant, then this can be viewed as a success. Ideally the process should be as flexible as the conditions it has to operate under are variable. Batch or modular plant seem to make allowances. However conventional activated sludge plant as described in the case history (6.10.4) remain an economic option where conditions allow and biological treatment for nutrient removal is the most sustainable option where appropriate. As the requirements for increased nutrient removal and compliance with UWWTD standards for coastal areas are implemented, a better understanding of the effects this will have on sludge settleability will become apparent. Correctly applying this knowledge represents the next challenge for water engineers and scientists.

6.11 USA (T. GLEN DAIGGER AND D. JENKINS)

6.11.1 General situations

Each year, at the beginning of the US Congressional Session, the President of the United States delivers a "State of the Union" message to a joint assembly of the Senate and the House of Representatives. In this address the President gives his (her) assessment of the condition of the United States and presents a summary of proposed programs. Since more weighty matters than activated sludge solids separation occupy the national agenda, we do not believe that this topic has ever been addressed in this forum. So it is left to us to report on the current state of this field in the United States.

Happily we can report that the state of the field is good!! The advances in the understanding of the population dynamics of the microorganisms that influence activated sludge solids separation that have taken place through research over the 27 years since Eikelboom (1973) devised a filamentous organism "identification" key have had a great influence on the design and operation of the activated sludge process in this country. Assessment and control of what is often known here as "sludge quality" are firmly established in professional wastewater treatment practice. Microscopic characterization of activated sludge is routinely practiced at many wastewater treatment plants, and it forms an integral part of troubleshooting activated sludge operating problems. Activated sludge design often incorporates features that discourage filamentous organism overgrowth (selectors, RAS chlorination) and encourage bioflocculation (secondary clarifier flocculating center wells). Many specialized courses are offered every year throughout the country on various aspects of activated sludge population assessment and control. Thus the "culture" of population dynamics has been embraced by the profession. This is not to say that all activated sludge solids separation problems have been resolved. It does mean, however, that when they occur there are a series of rational investigative and remedial tools available to address them.

6.11.2 Control of filaments in activated sludge systems

Our assessment of the current situation is that design and operating methods are available to address most of the filamentous bulking problems in activated sludge. The filamentous organisms most frequently causing bulking in activated sludge in the USA in 1984 are summarized in Table 6.24 (Richard et al 1981, Strom and Jenkins, 1984). Unfortunately no more recent comprehensive summaries have been conducted so that it is not possible to say whether Table 6.24 accurately represents the current situation. Control methods are available for all of these organisms as indicated in Table 6.24. Note that Table 6.24 does not contain "use of Cl_2" as a control method because it is universally successful.

Table 6.24. Summary of Most Frequently Encountered Filamentous Microorganisms in the USA in 1984, and Control Methods

Filamentous Organism	Control Method
type 1701/*S. natans*	raise DO, lower F/M
	anoxic, anaerobic selector
type 021N/*Thiothrix* spp	anoxic selector
	nutrient (N/P) supplementation
	remove SH^-
Haliscomenobacter hydrossis	raise DO, lower F/M
types 0041 and 0675	decrease MCRT
	anoxic/anaerobic selector (?)
M. parvicella	eliminate anoxic/anaerobic zones
	increase aerobic zone DO
	decrease MCRT
type 0092	decrease MCRT
	eliminate anoxic/anaerobic zones

Reflecting the adoption of filamentous organism control methods by the practicing profession, most US articles on bulking have dealt with field experience and case histories of design and performance. Table 6.25 provides a summary of publications on filament controls presented over the past years in USA technical journals. These publications demonstrate that tools are available to control excessive filament growths in essentially any activated sludge system.

An interesting exchange on the "success" of bulking control using selectors is presented in a discussion by Parker et al. (1998) of a paper by Martens and Daigger (1997), and the closure by Martens and Daigger (1998). The discussers argued that a good settling activated sludge should be defined as having an SVI of <100 mL/g for 90% of the time (rather than <150 mL/g), and that this value should be the target for plants with selectors if it is desired to achieve the greatest benefit from selector installation. They further suggested that a plant-scale research program be mounted to determine the SVI that can be achieved by various types of selector (aerobic, anoxic, anaerobic) and control measures (such as controls of MCRT, recycle rate and OUR) that might improve their performance. In their response, the authors demonstrated that such performance is possible with a properly designed and operated selector. Further research would serve to better define the design and operating parameters that that will produce such excellent settling sludges.

A recent development is the emergence of computerized filament identification keys. One example of these in the USA is the "Fil-identpro™", developed by The Stover Group (Stover & Associates, Inc., ANON). We may expect to see further developments like this in the future.

Table 6.25 Summary of Recent Filament Control Experience in the USA. And this is as long as the table can get if it goes to the full width

Filaments Studied	Filament Control Methods Investigated	Results	Reference
Sphaerotilus natans	Completely mixed anoxic selector studied at bench-scale using actual wastewater. Existing plant experienced severe bulking due to periodic discharges of readily biodegradable substrate.	Completely mixed anoxic selector effective at anoxic HRTs ranging from 17 to 164 min in system with total HRT (aeration basin plus clarifier) of 17 h and a total ST of 6 d.	Murakami and Babcock, 1998
Nocardioforms	Addition of 0.5 mg/L of a polyacrylamide cationic polymer to a full-scale WWTP with significant foam trapping characteristics.	Aeration basic foam coverage reduced significantly, and mixed liquor Nocardioform levels reduced to a lesser extent.	Starr, et al., 1996, Shao, et al., 1997
Nocardioform, Type 021N	Pilot-scale unit consisting of 4-stage anaerobic selector (25 % of total bioreactor volume), followed by 3-stage high purity oxygen bioreactor. Full-scale plant consisting of single stage anaerobic selector (one-sixth of total bioreactor volume) followed by 5-stage high purity oxygen bioreactor.	Effective filament control achieved in pilot-scale unit at 50 min anaerobic HRT, 2.4 d total MCRT, pH 6.3, temperature 20 °C. Effective filament control achieved in full-scale unit at 17 to 24 min anaerobic HRT, 3.5 d total MCRT, temperature = 20 °C. Filament control correlated with EBPR.	Fainsod, et al., 1996; Fainsod, et al., 1999
Nostocoida limicola II, Type 0041	The performance of anoxic selectors at 4 full-scale WWTPs were investigated.	Two factors were found necessary for effective filament control. One was an anoxic F/M of 0.8 to 1.2 kg BOD_5/kg MLSS, d, or an anoxic SRT of 1 d, for temperatures above 18°C and 0.7 to 1.0 kg BOD_5/kg MLSSd, or an anxic SRT of 1.4 d, for lower temperatures. The second was an aerobic SRT no more than 12 d.	Marten and Daigger, 1995; Marten and Daigger, 1997
Thiothrix II, *M. parvicella*, Nocardioforms, Types 0675 and 0092	Control of temperature of full-scale 2-stage carbon removal/nitrification activated sludge system treating pharmaceutical wastewater	Excessive filament growth observed at bioreactor temperatures of $\geq$ 40 °C, while filaments controlled at temperatures of 38 °C or below.	Bastain, et al., 1999
M. parvicella, Types 0092, 0041, and 0675	RAS Chlorination	Dosing at 2 g Cl_2/kg MLVSS, controlled SVI without adversely affecting N removal and with only modest effect on P removal. Higher Cl_2 doses controlled SVI but adversely affected N&P removal.	Ward, et al., 1999

cont'd

Table 6.25 cont'd

Filaments Studied	Filament Control Methods Investigated	Results	Reference
types 0041, and 0675,	Nutrient addition, intermittent aeration, aerobic selectors, anoxic selectors to treat municipal and food processing wastewater.	Intermittent aeration was effective, but operating problems were encountered due to use of gravity sludge recycle from the clarifier. Nutrient addition was marginally effective. A completely anoxic selector, operating at an F/M of 1.0 to 1.3 kg BOD_5/kg MLSS d was effective. Complete denitrification led to better SVI control. Aerobic selectors were most effective. Aerobic selector performance improved with increasing selector F/M ratio over the range of 0.1 to 1.4 kg BOD_5/kg MLSS d.	Nakhla and Lugowski, 1999
Not identified.	Aerobic and anaerobic selectors to treat cheese processing wastewater.	Single, completely mixed selector operating at F/M of 3 to 5 provided control, but severe foaming occurred due to low MLSS concentration. A single, completely mixed anaerobic selector operating at an F/M of 2.6 provided effective control of filaments.	Rein and Paulson, 1997
Not Identified	Changed supplemental ammonia dosing strategy for N deficient wastewater to ratio strategy.	Effective in reducing SVI from 145 to 117 mL/g and reduced filament abundance.	Barber, et al., 1997
Type 021N (Predominant) and Types 0041 and 1701, *Thiothrix* II and *S. natans*	Anaerobic selectors and sludge chlorination of long HRT (3.7 d) activated sludge system treating ceiling tile wastewater.	Chlorine was not effective when applied to the RAS, but was effective at doses between 1 and 3 g Cl_2/kg MLSS d when applied to the aeration basin. An anaerobic selector operating at an F/M of 14.3 kg BOD_5/kg MLSS d was partially effective.	Adey and McClintock, 1996
Nostocoida limicola II (Predominant)	Staged aerobic selector (three equal sized compartments) followed by completely mixed aerobic bioreactor to treat readily biodegradable industrial wastewater. Periodic sludge chlorination.	Aerobic selectors operating at first stage F/M values of about 10 kg COD/kg MLSS d and overall F/M values of about 3 kg COD/kg MLSS, d provided effective SVI control. An extended start-up was required for the sludge to acclimate and to acquire sufficient substrate storage capacity. Sludge chlorination was practiced periodically to maintain stirred SVI values below 100 mL/g.	Boe, et al., 1996
Nocardioforms	Classifying selector consisting of the addition of 0.5 mg/L of readily biodegradable nonionic surfactant and an aerated section with foam removal.	Effective at laboratory and full scale at controlling Nocardioform foaming.	Pagilla, et al., 1996

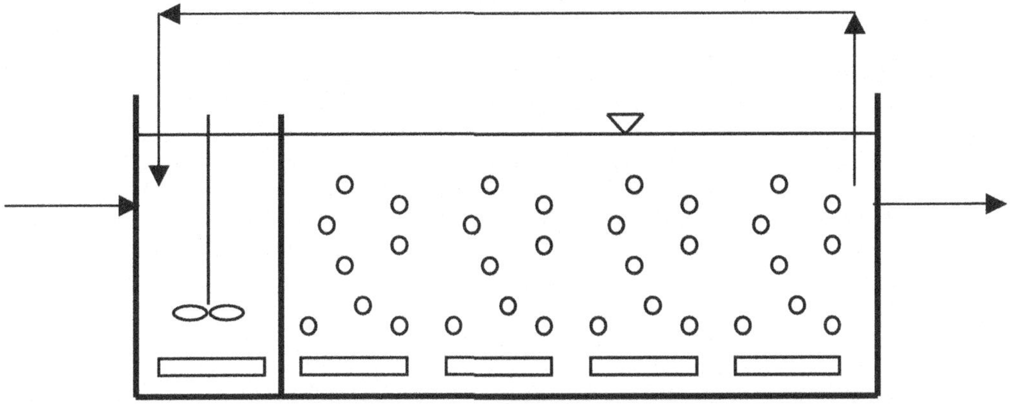

Fig. 6.27. Schematic of Tri-City WWTP Bioreactor Illustrating Anoxic Selector Operating Mode.

The Tri-City facility replaced two existing WWTP that both experienced severe bulking problems, which was one of the reasons that the anoxic selector features described above were incorporated into the facility design. In operation since 1986, facility operation has demonstrated the effectiveness of the anoxic selector. The facility has been operated in both the conventional and step feed activated sludge modes, resulting in sludge bulking as evidenced by SVI values well over 200 mL/g. In contrast, an SVI consistently below 120 mL/g and generally less than 100 mL/g can be maintained when operated in the anoxic selector mode. In fact, the selector is so effective that filaments can be eliminated to the point where a sludge that is too fragile is produced. When this occurs a small amount of air to the anoxic selector to grow low DO filaments. These features allow activated sludge settling characteristics to be controlled to essentially any desired value.

6.11.3 Other solids separation problems

We all know that filamentous bulking is only one of the microbially-based activated sludge solids separation problems. It is for the others, especially foaming, non-filamentous (or "viscous" or "zoogloeal") bulking and dispersed growth that advances in our understanding are needed to produce significant reliable control measures.

6.11.4 Foaming

In the USA, bacterial foaming of activated sludge is largely due to nocardioforms and to a lesser extent to *M. parvicella.*. With the increasing use of BNR the incidence of *M. parvicella* foaming may increase.

A relatively complete tool kit for solving nocardioform foaming problems has been developed, but the fundamental understanding of how some of these control measures work is not yet available (Table 6.26). We have found that control measures such as "washout" and anoxic or anaerobic selectors are universally successful when coupled with methods to control foam trapping. In contrast, systems with effective selectors and foam trapping characteristics may continue to be plagued with excessive foaming. Thus, attention to the details of the physical configuration of the facilities is required during design to minimize foam trapping. The prudent designer will also provide remedial foam control measures, such as surface chlorine sprays and/or cationic polymer addition capabilities.

Table 6.26. Nocardioform foaming control methods and status

Control Measure	Result	Research Required
Remove Foam Traps	Allows foam to escape. Prevents foam accumulation.	Does a foam trapping system influence nocardioform growth habit? Does a foam layer reseed mixed liquor?
Selective Foam Wasting	Selectively removes foam and associated organisms from process, thereby minimizing foaming	How selectively can nocardioforms be removed from process? Can other organisms (such as nitrifers) be enriched in mixed liquor?
Eliminate Foam Recycle	Prevents reseeding of activated sludge	None
Surface Cl_2 spray	Kills surface accumulations rich in nocardioforms	None
Reduce MCRT	Washout of nocardioforms	None
Aerobic, anoxic, anaerobic selectors	Deselects nocardioforms	None
Cationic polymer addition	Flocculates nocardioforms into floc? reducing exposure to bubbles	Proof of concept

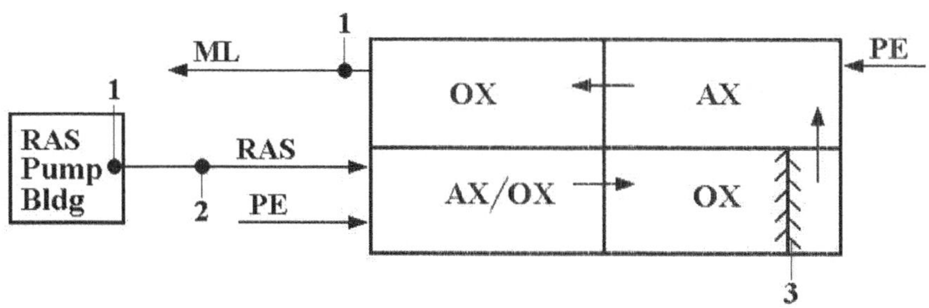

Fig. 6.28. San Jose/Santa Clara step WPCP Feed Aeration Basins. Some Nocardioform control features shown. 1: TSS meter; 2: Cationic polymer injection; 3: Cl_2 spray; AX: Anoxic; OX: Oxic; PE: primary effluent

The San Jose/Santa Clara WPCP has used several nocardioform foaming control methods (Ekster and Jenkins, 1999). This plant is a step feed nitrification/dentrification activated sludge system with basins that trap foam (Figure 6.28). Operating at MCRTs of between 5 to 7 days, it had been subject to nocardioform foaming incidents. Foaming control was first attempted by applying a fine mist spray of 3g/L NaOCl solution to the surface of the first aerobic zone using a shrouded spray bar (Figure 6.29). At doses of 0.5 to 1.0 mg Cl_2/L based on the wastewater flow rate this system effectively controlled nocardioform foam without altering bulk mixed liquid nocardioform organism levels, suggesting that it only kills nocardioforms at the surface of the basins. This method of nocardioform organism control did not compromise the BNR effluent in any way.

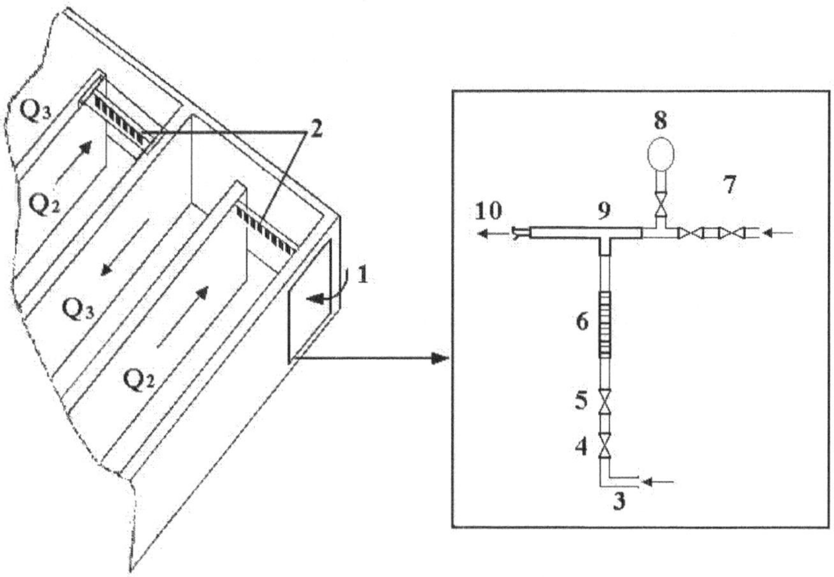

Fig. 6.29. Suface mist spray chlorination system for Nocardioform foam suppression; San Jose/Santa Clara WPCP; 1: Dose panel; 2: Hypalon should be closed at top and rear. At rear it extends a bit below liquid level to trap foam. Spray bar spaced spray nozzles; 3: From 12,5% NaOCl tank and pump; 4: Check valve; 5 Valve; 6: Rotameter; 7: Pressure regulator; 8: pressure gauge; 9: Eductor; 10: to spray bars

The second foam control method was to add 0.3 to 0.5 mg/L (based on combined wastewater and RAS flow) of a high charge catonic polymer to either the RAS or to the mixed liquor (Shao, et al., 1997). Foam disappeared from the aeration basins in under 10 days. These two foam control methods were unnecessary when it became apparent that after converting the system to automatic MCRT control (based on automatic on line MLSS and RAS SS measurements) there had been no nocardioform foaming incidents (even though the average MCRT was higher for automatic MCRT control than for manual control). Figure 6.30 illustrates this point and suggests that with manual MCRT control based on laboratory SS measurements, short periods of high MCRT exist that may allow nocardioforms to grow and become trapped in the aeration basins (Ekster and Jenkins, 1999).

6.11.5 Viscous Bulking and Dispersed Growth

It is in the area of viscous bulking and dispersed growth that the greatest research is needed. Viscous (non-filamentous zoogloeal) bulking refers to the production of slowly settling and poorly compacting activated sludge in the absence of large amounts of filamentous organisms. It is thought to be due to an overabundance of exocellular polymer that does not adhere well to the microbial surface.

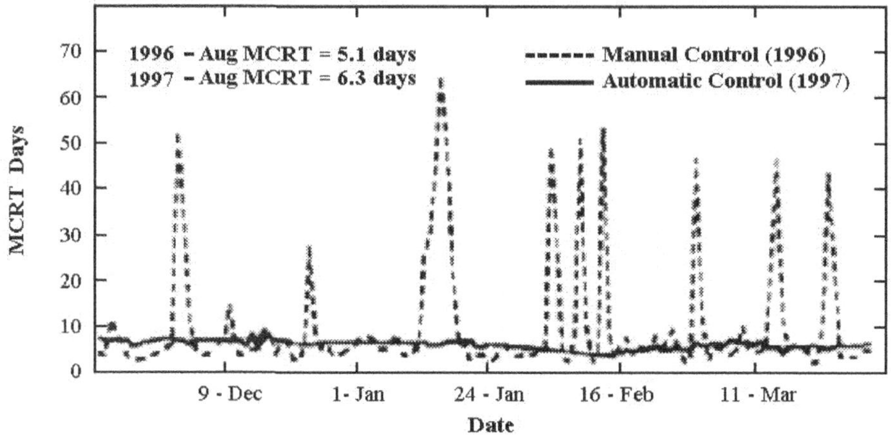

Fig. 6.30. MCRT Obtained with Manual Control (1996) and Automatic Control (1997) in the San Jose/Santa Clara, CA, BNR Plant.

Microbial exocellular polymers consist of a wide variety of compounds including polysaccharides, proteins, DNA, and phospholipids. In activated sludge, proteins and polysaccharides appear to be major components of the exocellular material (Higgins, 1995). Work by Novak's group (Higgins et al, 1999; Higgins and Novak, 1995, 1997a, 1997b, 1997c; Murthy and Novak, 1998; Murthy et al, 1998,; Novak et al 1998; Novak, 1997) has shown that a range of solids separation and dewatering properties are influenced by both the absolute and the relative amounts of monovalent cations (e.g., N_a^+, K^+, NH_4^+) and divalent cations (e.g., Ca^{2+}, Mg^{2+}). These properties include floc strength, settling rate, supernatant turbidity and soluble organic matter concentrations, filtration rates, capillary suction times, cake solids concentration, filtrate turbidity and soluble organic matter concentrations. A floc model (Figure 6.31) is proposed in which proteins (lectins) attached to the bacterial surfaces bridge between polysaccharides molecules with divalent cations. These cations also can bridge between anionic sites on adjacent polysaccharide molecules. A network is formed that binds the exocellular polysaccharides and proteins into the floc matrix.

When insufficient divalent cations and/or an overabundance of monovalent cations are present the protein/polysaccharide bridging is interfered with. The flocs loose their resistance to shear, resulting in dispersed particles that are difficult to settle and dewater, produce turbid supernatants (filtrates) and contain soluble polysaccharides that elevate the soluble organic matter content of the supernatant (filtrate). These types of effects seem to be most critical in solids dewatering and concentrating processes such as sludge filtration and centrifugation where particulate concentrations are the highest. However, many situations exist where these effects are important in the activated sludge process itself. These include the treatment of highly saline industrial wastewaters (especially where the salt derives from the use of NaOH to neutralize an acidic wastewater prior to biological treatment) and aerobic and possibly anaerobic digestion (where NH_4^+ concentrations can become substantial).

Several full-scale case studies (Higgins and Novak, 1997b; Murthy et al, 1998) have been presented where increased divalent/monovalent cation ratio in a wastewater has improved activated sludge and sludge dewatering solids separation process performances.

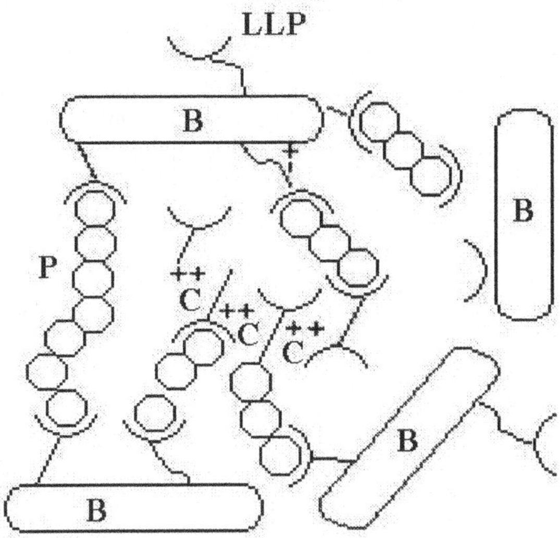

Fig. 6.31. Proposed roles of Polysaccharides and Proteins (Biopolymers) and Divalent Cations on Bioflocculation (Higgins, 1995). B: Bacteria; LLP: Lectin-like protein; C^{++}: Divalent cation; Polysaccharide

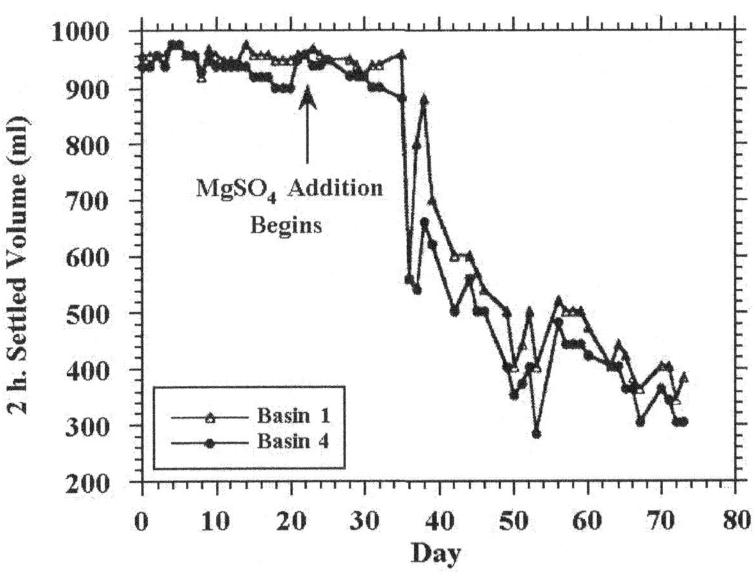

Fig. 6.32. Effect of $MgSO_4$ Addition on Activated Sludge Settling (Murthy, et al, 1998).

Murthy et al (1998) investigated the effect of divalent cation (Mg^{2+} as $MgSO_4$) addition on activated sludge settling at an industrial wastewater treatment plant which received a high COD concentration (10,000 mg/L, mostly acetate). The wastewater contained an average Na^+ influent concentration of 2,000 mg/L because it was neutralized (to pH 8.8) by NaOH addition. The effect of adding $MgSO_4$ to produce a feed concentration of 1 mM (24 mg/L

Mg^{2+}) is illustrated in Figure 6.32. Table 6.27 also shows that polymer usage for WAS thickening by gravity belt was significantly lower when $Mg(OH)_2$ was added.

Table 6.27 Comparison of Activated Sludge Operational Parameters During Month Long Trials with $Mg(OH)_2$ and NaOH Base Addition (Murthy, et al., 1998)

Parameter	$Mg(OH)_2$ Trial[a]	NaOH Trial[a]
Weekly Usage of Base, lb/wk	6,860	3,360
SVI, mL/g	170	200
Aeration Basin MLSS, mg/L	3,180	3,370
Digester MLSS, mg/L	17,600	16,800
Gravity Belt Polymer Usage, lb/wk	150	580

[a] Values represent monthly averages, unless otherwise specified.

6.11.6 Conclusions

The "culture" of activated sludge population dynamics and the concept of sludge quality are well established in the environmental engineering profession in the USA Filament and floc characterization of activated sludge systems for process control and problem diagnosis and resolution are becoming common. Rational design and operating methods are available and are being used in practice to prevent and control filamentous bulking and bacterial foaming. The sludge quality problems of "viscous" bulking and dispersed growth require further investigation.

6.12 REFERENCES

Adey A.H., McClintock S.A. (1996). "Control of Filamentous Bulking in a Complete-Mix Activated Sludge Process Treating High-Strength Industrial Wastewater " *Proceedings of the Water Environment Federation 69th Annual Conference & Exposition.* Volume 3. Remediation of Soil and Groundwater Industrial Wastes, 395-404.

Albertson O.E. (1991). Bulking sludge control: progress practice and problems *Water Sci & Technol*, **23**(4-6), 835-847.

Amann R., Ludwig W., Schleifer K.H. (1995). Phylogenetic identification and *in situ* detection of individual microbial cells without cultivation. *Microbiol Rev*, **59**, 143-169.

Andreadakis A. (1993). Physical and Chemical Properties of Activated Sludge. *Water Res*, **27**, 615-627.

Andreadakis A. (1998). Sludge Bulking in Treatment Plants in Greece *Final NTUA report to the Greek Secretariat for Research* (in Greek).

Andreasen K., Sigvardsen L. (1996). Experiences with sludge settleability in different process alternatives for nutrient removal. *Water Sci & Technol*, **33**(12), 137-146.

Andreasen K., Nielsen P.H. (2000). Growth of *Microthrix parvicella* in nutrient removal activated sludge plants - studies of in situ physiology. *Water Res*, **34**, 1559-1569.

Andreasen K., Nielsen P.H. (1997). Application of microautoradiography to study substrate uptake by filamentous microorganisms in activated sludge. *Appl Environ Microbiol*, **63**, 3662-3668.

Andreasen K., Nielsen P.H. (1998). *In situ* characterization of substrate uptake by *Microthrix parvicella* using microautoradiography. *Water Sci & Technol*, **37**(4-5), 19-26.

Arden E., Lockett W.T. 1914-1915. Experiments on the oxidation of sewage without the aid of filters – J Soc Chem Ind Lous, 33-34, part I II and III.

Ardem E. Lockett W.T. (1914). Experiments on the oxidation of sewage without the aid of filters. *J Soc Chem Ind*, **33**, 523-534.

Badoer S., Cuomo M., Miana P., Fabris M., Rossetti S., Tandoi V. Biological scum by *Microthrix parvicella* in a large wastewater treatment plant in Northern Italy: causes and solutions. Presented at IWA Regional Conference "Water key to sustainable development in Africa", Cape Town, 14-17 September 2003.

Barber J.B., Bullard C.M., Meyers Jr A. J.. "Nitrogen Supplementation Strategy for Variable-Strength Industrial Wastewater". *Proceedings of the Water Environment Federation 70th Annual Conference & Exposition.*

Volume 3, Remediation of Soil & Groundwater and Industrial Issues & Treatment Technology. Alexandria (VA), 1997, 551-57.

Bargna R., Terzi C., Ramadori R., (1993). Bulking – foaming control in the Varese – Olona plant. In Prevention and Control of Bulking Activated Sludge. 139-140. Jenkins D., Ramadori R., Cingolani L., Eds. Centro Studi Provincia di Perugia, Perugia, Italy.

Bastain O.J., Parks J.A., Bordacs K., Philp J.: "Effect of High Temperature on Filament Growth and Nitrifier Activity in an Activated Sludge Process: A Pilot Study ". *Proceedings of the Water Environment Federation 72nd Annual Conference & Exposition.* Water Environment Federation, Alexandria, (VA), CD-ROM 1999.

Beccari M., Ramadori R., Rozzi A., Tandoi V. (1995). Trattamenti biologici delle acque. In: Manuale di conduzione degli impianti di depurazione delle acque. Passino R. Ed., Zanichelli/Esac, Bologna.

Blackall L.L., Harbers A.E., Greenfield A.F., Hayward A.C. (1988). Actinomycete scum problem in Australian activated sludge plants. *Water Sci & Technol*, **20**(1-2).

Blackall L.L., Seviour E.M., Bradford D., Stratton H., Cunningham M., Hugenholtz P., Seviour R. (1996). Towards understanding the taxonomy of some of the filamentous bacteria causing bulking and foaming in activated sludge plants. *Water Sci & Technol*, **34**(5-6), 137-144.

Blackall L.L. (1986). Actinomycete scum problems in activated sludge plants. PhD Thesis, University of Queensland.

Blackall L.L., Harbers A.E., Greenfield P.F., Hayward A.C., (1991). Foaming in activated sludge plants: a survey in Queensland Australia and an evaluation of some control strategies. *Water Res*, **25**, 313-317.

Blackall L.L., Parlett J.H., Hayward A.C., Minnikin D.E., Greenfield P.F., Harbers A.E., (1989). *Nocardia pinensis* sp. Nov. and actinomycete found in activated sludge foams in Australia. *J Gen Microbiol*, **135**, 1547-1558.

Blackall L.L., Seviour E.M., Bradford D., Stratton H.M., Cunningham M.A., Hugenholtz P., Seviour R.J. (1996). Towards understanding the taxonomy of some of the filamentous bacteria causing bulking and foaming in activated sludge plants. *Water Sci & Technol*, **34**(5-6), 137-144.

Blackall L.L., Seviour E.M., Cunningham M.A., Seviour R.J., Hugenholtz P. (1994). Microthrix parvicella is a novel, deep branching member of the actinomycetes subphylum. Syst Appl Microbiol, **17**, 513-518.

Blackbeard J.R., Ekama G.A., Marais G.v.R. (1986). A survey of filamentous bulking and foaming in activated sludge plants in South Africa. *Wat Poll Control*, **1**, 90-100.

Blackbeard J.R., Ekama G.A., Marais G.v.R. (1986). A survey of bulking and foaming in activated sludge plants in South Africa. *Wat Pollut Control*, **95** (1) 90-100.

Blackbeard J.R., Gabb D., Ekema G.A., Marais G.v.R. (1988). Identification of filamentous organisms in nutrient removal activated sludge plants in South Africa. *Wat SA*, **14**, 29-34.

Boe R.S., Givens D., Jenkins D., Daigger G.T.: "Use of Selectors for Bulking Control in an Industrial Wastewater Activated Sludge Treatment Plant". Proceedings of the Water Environment Federation 69th Annual Conference & Exposition, Volume 3, Remediation of Soil and Groundwater Industrial Wastes, 1996, 487-497.

Borain G (1998). Personal communication, Umgeni Water.

Bradford D. (1998). Molecular biological studies of filamentous bacteria associated with activated sludge bulking and foaming. PhD Thesis, University of Queensland.

Bradford D., Hugenholtz P., Seviour E.M., Cunningham M.A., Stratton H.M., Seviour R.J., Blackall L.L. (1996). 16S rRNA analysis obtained from gram negative filamentous bacteria micromanipulated from activated sludge. *Syst Appl Microbiol*, **19**, 334-343.

Bux F. Kasan H.C. (1994). A microbiological survey of ten activated sludge plants. *Water SA*, **20**, 61-72

Casey T.G., Wentzel M.C., Loewenthal R.E., Ekama G.A., Marais G.v.R. (1992). An hypothesis for the cause of low F/M filaments bulking in nutrient removal activated sludge systems. *Water Res*, **26**, 867-869.

Casey T.G., Ekama G.A., Wentzel M.C., Marais G.v.R. (1992b). Causes and control of low F/M filamentous bulking in nutrient removal activated sludge systems. *Proc IAWQ Int Workshop on the prevention and control of bulking activated sludge*. Perugia Italy, 22-23 June, Jenkins D., Ramadori R., Cingolani L., Eds. 77-96.

Casey T.G., Ekama G.A., Wentzel M.C., Marais G.v.R. (1995). Filamentous organism bulking in nutrient removal activated sludge systems. A historical overview of causes and control. *Water SA*, **21**, 231-238.

Casey T.G., Wentzel M.C., Ekama G.A. (1999). Filamentous organism bulking in nutrient removal activated sludge systems Paper 11: A biochemical/microbiological model for proliferation of anoxic-aerobic (AA) filamentous organisms. *Water SA, 25*, 443-452.

Casey T.G., Wentzel M.C., Loewenthal R.E., Ekama Ga Marais GvR (1992a). A hypothesis for the cause of low F/M filament bulking in nutrient removal activated sludge systems *Water Res*, **26**, 867-869

Casey Tg Wentzel Mc Loewenthal Re Ekama G.A., Marais G.v.R. (1994). A hypothesis for the cause and control of anoxic-aerobic (AA) filamentous bulking in nutrient removal activated sludge systems. *Water Sci & Technol*, **29**(7), 203-212.

Cesaro G. (1997). Utilizzo di un selettore anossico nella lotta al bulking da Type 021N e *Thiothrix*. In: La Gestione degli Impianti di Depurazione delle Acque di Scarico. Esperienze Nazionali a Confronto, 43-49, Centro Studi Provincia di Perugia, Perugia Italy.

Chambers B., Tomlinson E.J., 1982: Bulking of activated sludge. Presentative and remedial methods, Chichester,: Ellis Horword Pub., 279.

Chartered Institution of Water and Environmental Management: Handbook of UK Wastewater Practice Activated Sludge. CIWEM Publications, Lavenham, UK, 1997.

Chiesa C., Irvine R.L., (1985). Growth and Control of filamentous microbes in activated sludge: an integrated hypothesis. *Water Res*, **19**, 471-479.

Chudoba J., Ottova V., Madera V. (1973). Control of activated sludge bulking part I and II. *Water Res,* **7**, 1163-1182 and 1389-1406.

Chudoba J., Ottova V., Madera V., (1973). Control of activated sludge filamentous bulking: I Effect of the hydaulic regime or degree of mixing in an aeration tank. *Water Res*, **7**, 1163-1182.

Chun J., Blackall L.L., Kang S.O., Hah Y.C., Goodfellow M. (1997). A proposal to reclassify *Nocardia pinensis*, Blackall *et al*, as *Skermania piniformis*, gen. nov. comb. Nov., *Int J Syst Bacteriol*, **47**, 127-131.

Cingolani L., Ciccarelli E., (1996). Depurazione Biologica a Fanghi Attivi; I microrganismi responsabili di bulking e schiume biologiche. Vol I, Centro Studi Provincia di Perugia, Perugia Italy.

Cingolani L., Ramadori R. (1992). Il controllo del bulking negli impianti a fanghi attivi. Centro Studi Provincia di Perugia, Perugia Italy.

Cooper C.W. (1998). Personal communication. Greater *Durban Metropolitan Council.*

Cyrus Z., Sladka A. (1970). Several Interesting Organisms Present in Activated Sludge. *Hydrobiologia*, **35**(3/4), 383 – 396.

Daigger G.T., Nicholson G.A. (1990). Performance of Four Full-Scale Nitrifying Wastewater Treatment Plants Incorporating Selectors. *Res J Water Pollut Control Fed,* **62**, 676-683.

Davoli C., Guglielmi L. (1998). Utilizzo della clorazione nella lotta al bulking filamentoso. Technical Report AGAC Ed., Reggio Emilia, December 1998.

Davoli D., Madoni P., Collini G., Pergetti M. (1997). Approccio sperimentale al controllo del fenomeno di foaming biologico in un impianto di depurazione a fanghi attivi. *Acqua e Aria*, (4), 77-83.

Davoli D., Madoni P., Guglielmi L., Pergetti M., Barilli S. (1997). Testing the effect of selectors in the control of bulking and foaming in full scale activated-sludge plants. *Water Sci & Technol*, **46**(1-2), 495-498.

de los Reyes F.L. III, Raskin L.: "Population Dynamics of filamentous Formaing in Activated Sludge: Toward an Integrated Theory". *Proceedings of the Water Environment Federation 71st Annual Conference & Exposition*, Volume 1, Wastewater Treatment Research and Municipal Wastewater Treatment. Water Environment Federation, Alexandria, VA, 1998, 281-291.

Dhaliwal B.S. (1979). *Nocardia amarae* and activated sludge foaming. J *Wat Pollut Control Fed*, **51**, 344-350.

Di Marino R., Donadio F., Verdiani B.: "Alcune esperienze sul controllo del bulking filamentoso e foaming. Indagini ed interventi proposti". Il bulking filamentoso: controllo e gestione. AGAC Ed., Reggio Emilia, Italy, 5 Maggio 1992.

Dohanyos M., Grau P., Chudoba J. (1971). Kinetic assessment of glucose removal and saccharide accumulation capacities in activated sludge. *Wat Poll Res*, Jenkins S.H. Ed., II-3/1, II-3/7, Pergamon Press, Oxford UK .

Eikelboom D.H. (1975). Filamentous organisms observed in activated sludge. *Water Res*, **9**, 365-388.

Eikelboom D.H., van Buijsen H.J.J. (1981). *Microscopic Sludge Investigation Manual*, IMG-TNO, Report A94a, Delft NL.

Eikelboom D.H. (1982). Biosorption and prevention of bulking sludge by means of a high floc loading. In: Bulking of activated sludge, Chambers B., Tomlinson E.J. Eds, 90-104, Ellis Horwood WRC Ltd, Chichester UK.

Eikelboom D.H. (1987). Bulking sludge control manual (in Dutch). *TNO Report R87/358.*

Eikelboom D.H. (1991). Schuim-en drijflaagvorming op zuiveringsinrichteingen. TNO Milieu en energie report 91/305 Delft, NL.

Eikelboom D.H. (1993). Effect of N and P removal on the settling properties of activated sludge (in Dutch). *TNO Report R93/074.*

Eilkelboom, D.H.(1997): Control of *Microthrix parvicella* by addition of PAX 14. TNO Report R 97/305.

Eikelboom D.H. (2000). Process control of activated sludge plants by microscopic investigation. Handbook IWA Publishing.

Eikelboom D.H., van Buijsen H.J.J. (1983). Microscopic sludge investigation manual 2nd ed., TNO Research Institute for Environmental Hygiene Delft the Netherlands, 1-81.

Eikelboom D.H., Andreasen K. (1995). Survey of the filamentous population in nutrient removal plants in four European Countries. TNO-Report MW-R 95/090, Delft, NL .

Eikelboom D.H., Andreadakis, A., and Andreasen K. (1998). Survey of the filamentous population in nutrient removal plants in four European Countries. *Water Sci & Technol*, **37** (4-5), 281-289.

Eikelboom D.H., Grovenstein J. (1998). Control of bulking in full scale plant by addition of talc (PE 8418). *Water Sci & Technol*, **37**(4-5), 297-303.

Eikelboom D.H., Geurkink B. (2002). Filamentous micro-organisms observed in industrial activated sludge plants *Water Sci & Technol* **46**(1-2), 535-542.

Ekama G.A. (1998). Personal communication. *Water Research Group, Dept of Civil Eng Univ of Cape Town.*

Ekama G.A., Wentzel M.C., Casey T.G., Marais G.v.R. (1996a). Filamentous organism bulking in nutrient removal activated sludge systems. Stimulation of the selector effect under anoxic conditions. *Water SA*, 22, 119-126.

Ekama Ga Wentzel Mc Casey Tg Marais GvR (1996b). Filamentous organism bulking in nutrient removal activated sludge systems Paper 6: Review evaluation and consolidation of results *Water SA* **22**(2) 147-152.

Ekster A., Jenkins D.: "*Nocardia* Foaming Control in a BNR Plant". *Proceedings of the Water Environment Federation, 72nd Annual Conference & Exposition*, Water Environment Federation Alexandria VA CD-ROM 1999.

Erhart R., Bradford D., Seviour R.J., Amann R., Blackall L.L. (1997). Development and use of fluorescent in situ hybridization probes for the detection and identification of "*Microthrix parvicella*" in activated sludge. *Syst Appl Microbiol*, **20**, 310-318.

Fainsod A., Pagilla K.R., Jenkins D., Pitt P.A., Mamais D. (1999). The Effect of Selectors on Nocardioform Organism Growth in Activated Sludge. *Water Environ Research*, **71**, 1151-1157.

Fainsod A., Pagilla K.R., Jenkins D., Pitt P.A., Mamais D. "The Effect of Anaerobic Selectors on *Nocardia* Growth in Activated Sludge " *Proceedings of the Water Environment Federation 69st Annual Conference & Exposition*. Volume 1, Wastewater Treatment Research and Municipal Wastewater Treatment. Water Environment Federation, Alexandria (VA), 1996, 189-200.

Falco R.J., van Hoeck M., Richard M., "Viscous Poly- Saccharide Production During Treatment of Industrial Wastewater". *Proceedings of the Water Environment Federation 72nd Annual Conference & Exposition.* Water Environment Federation, Alexandria (VA), 1999, CD-ROM.

Fisher M.J., Boyle W.C. (1999). Effect of anaerobic and anoxic selectors on oxygen transfer in wastewater. *Water Environ Res*, **71**, 84-93.

Foot R., Forster C. (1997). The operation of a selector for the control of foam-forming bacteria in activated sludge. *Environ Technol*, **18**, 237-241.

Foot R., Robinson M. (2003). Activated sludge bulking and foaming – microbes and myths. *In "'The Handbook of Water and Wastewater Microbiology".* Duncan Mara and Nigel Horan Eds., Academic Press, 32, 525-544.

Foot R., Robinson M., Forster C. (1993). Operational aspects of three 'selectors' in relation to aeration tank ecology and stable foam formation. *J Instn Wat & Envir Mangt*, 7, 304-309.

Foot R., Robinson M., Forster C. (1994). Systematic activated sludge bulking and control. *Water Sci & Technol*, **29**(7), 213-220.

Gabb D.M.D., Ekama G.A., Jenkins D., Wentzel M.C., Casey T.G., Marais G.v.R. (1996a). Filamentous organism bulking in nutrient removal activated sludge systems. Paper 4: System configurations and operating conditions to develop low F/M bulking sludges at laboratory-scale. *Water SA*, **22**, 127-138.

Gabb D.M.D., Ekama G.A., Jenkins D., Wentzel M.C., Casey T.G., Marais G.v.R (1996b). Filamentous organism bulking in nutrient removal activated sludge systems. Paper 5: Experimental examination of aerobic selectors in anoxic-aerobic systems. *Water SA*, 22, 139-146.

Godley A., Junkermann H., Knab V. (1998). Application of intermittent denitrification to the treatment of refinery wastewater. *Indust Effluent Technol, 95-102.*

Goodfellow M., Chun J., Stubbs S., Tobili A.S. (1994). Transfer of *Nocardia amarae*, Lechevalier & Lechevalier 1974, to the genus *Gordona* as *Gordona amarae*, comb. nov. *Lett Appl Microbiol*, **19**, 401-405.

Goodfellow M., Davenport R., Stainsby F.M., Curtis T.P. (1996). Actinomycete diversity associated with foaming in activated sludge plants. *J Indust Microbiol*, **17**, 268-280.

Gostic N. (1995). Activated sludge - filamentous bulking. *Water & Waste Treatment*, 38, 16-19.

Hao O.J., Strom P.F., Wu Y.C. (1988). A review of the rôle of *Nocardia* like filaments in activated sludge foaming. *Water SA*, **14**, 105-110.

Heide B.A., Pasveer A. (1974). Oxidation ditch: Prevention and control of filamentous bulking. *H_2O*, **7**, 373-377.

Henze M.P., Harremoes J., Jansen J., Arvin E. (1992). Wastewater Treatment: Biological and Chemical Processes. Springer Verlag, Berlin.

Higgins M.J. "The Interaction of Cations Proteins and Poly-Saccharides in the Settling and Dewatering of Activated Sludge", PhD, Dissertation Environmental Engineering, Virginia Polytechnic and State University, August 1995.

Higgins M.J., Novak J.T. (1997a). The Effect of Cations on the Settling and Dewatering of Activated Sludges: Laboratory Results. *Water Environ Res.* **69**, 215-224.

Higgins M.J., Novak J.T. (1997b) Dewatering and settling of activated sludges: the case for using cation analysis. *Water Environ Res*, **69**, 225-232.

Higgins M.J., Novak J.T. (1997c). Characterization of Exocellular Protein and Its Role in Bioflocculation. *J Environ Eng*, 123, 479-485.

Higgins M.J., Novak J.T. "The Effect of Cations on the Settling and Dewatering of Activated Sludges", *Proceedings of the Water Environment Federation, 68st Annual Conference & Exposition*, Volume 1, Wastewater Treatment Research and Municipal Wastewater Treatment, Water Environment Federation, Alexandria (VA) 1995 225-236.

Higgins M.J.,. Miller L.A, Sobeck D."Improving Industrial Wastewater Treatment Using Divalent Cation Addition", *Proceedings of the Water Environment Federation 72nd Annual Conference & Exposition*, Water Environment Federation, Alexandria (VA), CD-ROM 1999.

Hilditch J. (1998). Personal communication. *Greater Durban Metropolitan Council.*

Hiraoka M. Tsumura K. (1984). Suppression of actinomycete scum production: a case study at Senboku wastewater treatment plant, Japan. *Water Sci & Technol*, 16(10-11), 83-90 .

Istat. Gli impianti di depurazione delle acque reflue urbane. Anno 1993, Informazioni n 3, 1996.

Jacobsen B.N., Guildal T., Vestergaard J., Andreasen K., Sigvardsen L. (1997). Full scale experience with the control of *Microthrix parvicella. Proceedings 2nd Int Conf on Microorg in Activated Sludge and Biofilm Processes*, Berkeley (CA), 473-476.

Japan Sewage Works Agency (1983). Survey report on measures to control activated sludge bulking and foaming. Intermediate Report I (in Japanese).

Japan Sewage Works Agency (1984). Survey report on measures to control activated sludge bulking and foaming. Intermediate Report II (in Japanese).

Japan Sewage Works Agency (1985). Survey report on measures to control activated sludge bulking and foaming (in Japanese).

Japan Sewage Works Agency (1986). Survey report on measures to control activated sludge bulking and foaming (in Japanese).

Japan Sewage Works Agency (1994). Survey report on activated sludge separation problems (in Japanese).

Jenkins D., Richard M.G., Daigger G.T., (1993). *Manual on the Causes and Control of Activated Sludge Bulking and Foaming*, 2nd edition, Lewis Publishers, Bocca Raton (FL), USA.

Jenkins D., Richard M.G., Daigger G.T. (1986). Manual on the causes and control of activated sludge bulking and foaming Ridgeline Press, - Lafayette (CA), USA.

Jenkins D., Richard M.G., Daigger G.T. (1984). *Manual on the causes and control of activated sludge bulking and foaming*. Water Research Commission Ed., Pretoria, South Africa.

Johannes W.G., St Arnaud J. (1998). Personal communication, *Africon Engineering/Rustenburg Council.*

Kanagawa T., Kamagata Y., Yamada K., Kohno T. (1998). Diversity of Type 021N isolated from bulking sludges. *Proc 1st Symposium of Japan Society on Water Environment*, 205-206 (in Japanese).

Khan A.R., Foster C..F (1988). Biosurfactant production by *Rhodococcus rubra. Envir Technol Lett*, 9, 1349-1360.

King P.B. (1998). Personal communication, *Cape Metropolitan Council.*

Knight G.C., Seviour E.M., Seviour R.J., Soddell J.A., Lindrea K.C., Strachan W., DeGrey B., Bayly R.C. (1994). Development of the microbial community of a full scale biological nutrient removal activated sludge plant during start up. *Water Res*, **29**, 2085-2093.

Knoop S., Kunst S. (1998). Influence of temperature and sludge loading on activated sludge settling especially on *Microthrix parvicella. Water Sci & Technol*, **37**(4-5), 27-35.

Kocianova E., Foot R., Forster C. (1992). Physicochemical aspects of activated sludge in relation to stable foam formation. *J Instn Wat & Envir Mangt*, **6**, 342-350 .

Kohno T. (1987). Microbiological study on activated sludge bulking. PhD thesis, Yamanashi University

Kohno T (1988). Morphology physiology and nutrition of a sulfur oxidizing filamentous organisms isolated from bulking activated sludges. *Water Sci & Technol*, **20**(11-12), 241-247.

Kos M., Wanner J., Sorm I., Grau P. (1992). R-D-N Activated Sludge Process. *Water Sci & Technol,* **25**(4-5), 151–160.

Krhutkova O., Ruzickova I., Wanner J. (2001). Microbial evaluation of activated sludge and filamentous population at eight Czech nutrient removal activated sludge plants during year 2000. *Proc of the 3rd IWA International Specialised Conference on Microorganisms in Activated Sludge and Biofilm Processes*, 13–15 June Rome Italy, Tandoi V., Passino R., Blundo M.C. Eds..

Kristensen G.H., Jørgensen P.E., Nielsen P.H. (1994). Settling characteristics of activated sludge in Danish treatment plants with biological nutrient removal. *Water Sci & Technol*, **29**(7), 157-165.

Kroiß H. (1981). Adsorptions- und Speichervorgänge beim Belebungsverfahren (Adsorption and storage processes in activated sludge treatment in German). *GWF Wasser-Abwasser*, **19** (1) 101-122 .

Kroiß H., Fenz R., Prendl L. (1997). Behandlung von Industrieabwasser am Beispiel der Zuckerindustrie (Sugar industry wastewater treatment in German), *Wiener Mitt*, **141**, 187-217.

Kruit J., Hulsbeek J., Visser A. (2002). Bulking sludge solved?! *Water Sci & Technol*, **46**(1-2), 457-464.

Kunst S., Knoop S. (1996). Schaum in Faulbehältern (Foam in digesters in German). In: "Ökologie der Abwasserorganismen" (Ecology of wastewater organisms). Lemmer H., Griebe T., Flemming H.C. Eds, 273-289, Springer Verlag, Heidelberg, Germany.

Lakay M.T., Hulsman A., Ketley D., Warburton C., De Villiers M., Casey T.G., Wentzel M.C., Ekama G.A. (1999). Filamentous organism bulking in nutrient removal activated sludge systems – Paper 7: Exploratory experimental investigations. *Water SA*, 25, 383-396.

Lakay M.T., Wentzel M.C., Ekama G.A., Marais G.v.R. (1988). Bulking control with chlorination in a nutrient removal activated sludge system. *Water SA*, **19**, 35-42.

Lemmer H., Kroppenstedt R.M. (1984). Chemotaxonomy and physiology of some actinomycetes isolated from scumming activated sludge. *Syst Appl Microbiol*, **5**, 135-135.

Lemmer H. (1996). Biologische Ursachen von Schaum und Schwimmschlamm in Belebungsanlagen sowie mögliche Gegenmaßnahmen (Biological causes of scum and floated sludge and troubleshooting measures in German). In: "Ökologie der Abwasserorganismen (Ecology of wastewater organisms)". Lemmer H., Griebe T. Flemming H.C. Eds), 259 – 271, Springer Verlag Heidelberg, Germany.

Lemmer H., Eikelboom D., Kappeler J., Klein B., Kunst S., Matsché N., Popp W., Schön G., Wagner F., Wolfgramm J., Zander-Hauck S. (1998). Blähschlamm Schwimmschlamm und Schaum in Belebungsanlagen - Ursachen und Bekämpfung (Bulking floated sludge and scum in activated sludge plants – causes and troubleshooting measures, in German). Report of the ATV-DVWK task group KA 61 "Bulking scum biological additives", Korr Abw, **45**(10), 1959-1968 and Korr and Abw **45** (11) 2138.

Lemmer H., Lind G. (2000). Blähschlamm Schaum Schwimmschlamm Mikrobiologie und Gegenmaßnahmen (Sludge bulking scum floated sludge Microbiology and countermeasures in German). Hirthammer, München, Germany.

Lemmer H., Lind G., Müller E., Schade M. (2004). Biologische Charakterisierung und Bekämpfung weniger bekannter Schaumbakterien (Biological characterization and countermeasures against non-famous scum bacteria, in German). Final Report Project 02WA0121 "Schaum in Anlagen zur biologischen Abwasser- und Schlammbehandlung", German Federal Ministry of Education and Research BMBF Ed., LfW-Part 3, München, Germany.

Lemmer H., Lind G., Müller E., Schade M., Ziegelmayer B. (2000). Scum in activated sludge plants: Impact of non-filamentous and filamentous bacteria. *Acta Hydrochim Hydrobiol*, **28**, 34-40.

Lind G., Lemmer H. (1998). Biologische Charakterisierung von Schäumen in Belebungsanlagen Teil I, Bedeutung fädiger Belebtschlammbakterien (Biological characterization of scum in activated sludge, Part I. Impact of filamentous activated sludge bacteria, in German, *GWF Wasser-Abwasser*, **139**, 1-6.

Little C. (1998). Personal communication, *Ninham Shand (CT)*.

van Loosdrecht M.F., Brandse A., de Vries A.C. (1998). Upgrading of waste water treatment process for integrated nutrient removal. The BCFS process, *Water Sci & Technol*, **37**(9), 209-217.

Madoni P., Davoli D., Gibin G., (2000). Survey of filamentous microorganisms from bulking and foaming activated sludge plants in Italy. *Water Res*, 34, 1767-1772.

Madoni P., Davoli D., Guglielmi L., Stefanini C. (1996). Guida all'analisi microscopica del fango attivo. Atlante Fotografico, AGAC Reggio Emilia and Università degli Studi di Parma Eds.

Mamais D., Andreadakis A., Noutsopoulos C., Kalergis C. (1998). Causes of and control strategies for *Microthrix parvicella* bulking and foaming in nutrient removal activated sludge systems. *Water Sci & Technol*. **37**(4-5), 9-18.

Martens W.L., Daigger G.T. (1997). Full-scale evaluation of factors affecting the performance of anoxic selectors. *Water Environ Res*, **69**, 1272-1281.

Martens W.L., Daigger G.T. (1998). Closure to full-scale evaluation of factors affecting the performance of anoxic selectors. *Water Environ Res*, **70**, 1229-1231.

Martens W.L., Daigger G.T. (1995). Full-Scale Evaluation of Factors Affecting the Performance of Anoxic Selectors. *Proceedings of the Water Environment Federation 68st Annual Conference & Exposition.* Volume 1, 669-680, Wastewater Treatment Research and Municipal Wastewater Treatment, Water Environment Federation, Alexandria (VA).

Matsché N. (2003). Application of detergents to improve scum handling. Project Report, Technical University Vienna, Austria.

Matsché N. (1991). Erfahrungen mit der Anwendung von Kalk auf Abwasserreinigungsanlagen, in Österreich (Experiences with lime addition in wwtps in Austria, in German). Nachlese Symposium für Abwasser und Klärschlammbehandlung der Österreichischen Kalkindustrie, 13-15.

Matsché N., Kreuzinger N. (1998). Blähschlamm Schwimmschlamm Schaum - Ursachenerkennung und Bekämpfung (Bulking floated sludge scum – causes and countermeasures. in German). *Wiener Mitt*, **145**, 141-177.

Miana P., Grando L., Caravello G., Fabris M. (2002). *M. parvicella* foaming at the Fusina WWTP. *Water Sci & Technol*, **46**(1-2), 499-502.

Mino T. (1995) Survey on Fillamentous Microorganisms in Activated Sludge Processes in Bangkok Thailand, *Water Sci & Technol*, **31**(9), 193-202.

Molinari M., Canziani R. (1992). Analisi Microscopica dei fanghi attivati In Rimozione Biologica dell'azoto: impianti per reflui industriali, 193–228. Canziani R., Dubini M., Molinari M. Eds., Istituto per l'Ambiente, Milano.

Moryoussef A., and Trousselle C. (1998) Proceedings of international symposium "Waterways Rehabilitation in Urban Environments", 1- 6, Quebec, Canada.

Müller E. (2005). Bacteria and extracellular polymeric substances in activated sludge scum formation. PhD thesis, Technical University Munich, in preparation.

Müller E., Lind G., Lemmer H., Wilderer P.A. (2004). Biologische und chemische Charakterisierung von Bioschäumen (Biological and chemical characterization of scum, in German). Final Report, Project 02WA0121. "Schaum in Anlagen zur biologischen Abwasser und Schlammbehandlung", German Federal Ministry of Education and Research BMBF Eds., LfW-Part 1, München, Germany.

Murakami C., Babcock J. (1998). Effect of anoxic selector detention time and mixing rate on denitrification rate and control of *Sphaerotilus natans* bulking. *Proceedings of the Water Environment Federation, 71st Annual Conference & Exposition*, Wastewater Treatment Research and Municipal Wastewater Treatment, Volume 1, 89-98, Alexandria (VA), USA.

Murthy S.N., Novak J.T. (1998). Influence of cations on activated sludge effluent quality. *Proceedings of the Water Environment Federation, 71st Annual Conference & Exposition*, Volume 1, 309-324. Wastewater Treatment Research and Municipal Wastewater Treatment. Alexandria (VA), USA.

Murthy S.N., Novak J.T., De Haas R.D. (1998). Monitoring cations to predict and improve activated sludge settling and dewatering properties of industrial waters. *Water Sci & Technol*, **38**(3) 119-126.

Nakhla G., Lugowski A. (1999). Full-scale comparative performance assessment of selectors and intermittent aeration for filament control. *Proceedings of the Water Environment Federation, 72nd Annual Conference & Exposition*, CD-ROM, Alexandria (VA), USA.

Nielsen P.H., Andreasen K., Wagner M., Blackall L.L., Lemmer H., Seviour R.J. (1998). Variability of Type 021N in activated sludge as determined by in situ substrate uptake pattern and in situ hybridization with fluorescent rRNA targeted probes. *Water Sci & Technol* **37**(4-5), 423-430.

Novak J.T. (1997). The effect of nitrification on activated sludge settling properties. *Proceedings of the Water Environment Federation, 70st Annual Conference & Exposition*, Volume 1, 377-384, Research and Municipal Wastewater Treatment. Alexandria (VA), USA.

Novak J.T., Love N.G., Smith M.L., Wheeler E.R. (1998). The effect of cationic salt addition on the settling and dewatering properties of an industrial activated sludge. *Water Environ Res*, **70**, 984-996.

Osborn D.W. (1988). Sewage purification in South Africa – past and present. *Water SA*, 14, 139-151.

Pagilla K.R., Jenkins D., Kido W.H. (1996). *Nocardia* control in activated sludge by classifying selectors. *Water Environ Res*, **68**, 235-239.

Paris S., Lind G., Lemmer H., Wilderer P.A. (2004). Bekämpfung von Schwimmschlamm verursacht durch *Microthrix parvicella* (Troubleshooting scum caused by *M parvicella*, in German). Final Report Project 02WA0121, "Schaum in Anlagen zur biologischen Abwasser- und Schlammbehandlung", German Federal Ministry of Education and Research BMBF Ed., WGA, München, Germany.

Parker D., Ekama G.A., Melcer H., van Niekirk A. (1998). Discussion of full-scale evaluation of factors affecting the performance of anoxic selectors, by Marten W.L. and Daigger G.T.. *Water Environ Res*, **70**, 1225-1228.

Pasinetti E., Emondi V. (2005): Survey of activated sludge filamentous microorganisms from several North Italian Wastewater Treatment Plants. 6th International Conference and Exhibition, 131-137, Teplice, Czech Republic, ACE Ed.

Pipes W.O. (1978). Actinomycetes scum production in activated sludge processes. *J Wat Pollut Control Fed*, 50, 628-634.

Pitman A.R. (1996). Bulking foaming in BNR plants in Johannesburg: problems and solutions. *Water Sci & Technol*, **34**(3-4), 291-298.

Pitman A.R. (1985). Settling of nutrient removal activated sludges. *Water Sci & Technol*, **17**(4-5), 493-504.

Pitman A.R. (1991). Design considerations for nutrient removal activated sludge plants. *Water Sci & Technol*, **23**(5), 781-790.

Pitman A.R. (1994). Process design of new biological nutrient removal extensions to Northern Works Johannesburg. In *"Nutrient Removal from Wastewaters"*, 167-172. Horan N.J., Lowe P., Stentiford E.I. Eds., Technomic Publishing, Co Inc..

Pitman A.R. (1995). Practical experience with biological nutrient removal on full-scale wastewater treatment plants in South Africa. *Veroffentlichungen des Institutes fur Siedlungswasserwirtschaft und Abfalltechnik der Universitut Hannover*. Heft 92, Bio-P Hannover 95, Internationale Konferenz zur vermehrten biologischen Phosphorelimination, 6a-1- 6a-21.

Pitman A.R. (1996). Bulking and foaming in BNR plants in Johannesburg: Problems and solutions. *Water Sci & Technol*, **34**(3-4), 291-298.

Pitman A.R. (1997). Nutrient Removal: Experiences in Johannesburg. (Abridged). *JCIWEM*, 12, 338-345.

Prendl L. (1997). Ein Beitrag zu Verständnis und Anwendung aerober Selektoren für die Blähschlammvermeidung (Contribution to a better knowledge of aerobic selectors and their application as a means for troubleshooting bulking problems, in German). *Wiener Mitt*, **139.**

Pretorius W.A., Laubscher C.J.P. (1987). Control of biological scum in activated sludge by means of selective flotation. *Water Sci & Technol*, **19**(5-6), 1003-1011.

Pujol R. (1984). La zone de contact - premiers résultats. TSM, 4, 199-205.

Pujol R. (1991). Applied ecology to filamentous microorganisms in activated sludge: the key to bulking control. Approach to sewage treatment process. 157-161, Madoni P. Ed, Perugia.

Pujol R., Canler J.P. (1989). Le foisonnement des boues activées: situation du problème en France TSM, (1), 19-24.

Pujol R., Canler J.P. (1994). Contact zone: French practice with low F/M bulking control. *Water Sci & Technol*, 29(7), 221-228.

Pujol R., Duchene P., Schetrite S., Canler J,P, (1991). Biological foams in activated sludge plants: characterization and situation. *Water Res*, **25**, 1399-1404.

Raunkjær K.T., Jacobsen H., Nielsen P.H. (1994). Measurement of pools of protein carbohydrate and lipids in domestic wastewater. *Water Res*, **28**, 251-262.

Rein D.A., Paulson S. (1997). Selectors for high strength industrial wastewater treatment. *Proceedings of the Water Environment Federation 70th Annual Conference & Exposition.* Volume 3, 415-424. Remediation of Soil & Groundwater and Industrial Issues & Treatment Technology, Alexandria (VA), USA.

Richard M.G., Hao O.J., Shimiju G., Jenkins D. (1981). The isolation and characterization of filamentous microorganisms from activated sludge bulking. National Science Foundation Report, December.

Rossetti S., Carucci A., Rolle E. (1994). Survey on the occurrence of filamentous organisms in municipal wastewater treatment plants relating to their operating conditions. *Water Sci & Technol*, **29**(7), 305-308.

Saayman G.B., Schutte C.F., van Leeuwen J. (1996). The effect of chemical bulking control on biological nutrient removal in a full scale activated sludge plant *Water Sci & Technol*, **34**(3-4) 275-282.

Saayman G.B., Schutte C.F., van Leeuwen J. (1997). Full-scale chemical control of sludge bulking. *Water Research Commission Report* **328/1/97**, Water Research Commission Ed., Pretoria, South Africa.

Saayman G.B., van Leeuwen J. (1992). Full-scale bulking control by ozonation. The use of ozone in water and wastewater treatment. *Proc of the second International ozone conference.* Warmbaths, South Africa.

Sarfert F., Eikelboom D.H., Klein B., Kowalsky H., Lemmer H., Matsché N., Mudrack K., Popp W., Reinnarth G., Wagner F. (1988). Verhinderung und Bekämpfung von Blähschlamm und Schwimmschlamm (Prevention and troubleshooting sludge bulking and scum, in German). ATV task group 261 Report: Sludge bulking and countermeasures, *Korr Abw*, **35**, 152-164.

Schade M., Lemmer H. (2002). Charakterisierung von Fadenbakterien zur Bekämpfung von Schaum in Belebungsbecken (Characterization of filamentous bacteria for scum troubleshooting in activated sludge wwtps in German). Final report. Bayerisches Landesamt für Wasserwirtschaft Ed., Materialien Nr 104, München, Germany.

Schade M., Lemmer H. (2004). Enzymaktivitätsmessungen (Enzyme activity in scum in German). Final Report. Project 02WA0121 "Schaum in Anlagen zur biologischen Abwasser- und Schlammbehandlung". German Federal Ministry of Education and Research BMBF Ed., LfW-Part 2, München Germany, *Sci and Eng*, **20**, 1.

Seviour E.M., Blackall L.L., Christensson C., Hugenholtz P., Cunningham M.A., Bradford D., Stratton H.M., Seviour R.J. (1997). The filamentous morphotype Eikelboom Type 1863 is not a single genetic entity. *J Appl Microbiol*, **82**, 411-421.

Seviour E.M., Williams C.J., DeGrey B., Soddell J.A., Seviour R.J., Lindrea K.C. (1994). Studies on filamentous bacteria from Australian activated sludge plants. *Water Res*, **28**, 2335-2342.

Seviour E.M., Williams C.J., Seviour R.J., Soddell J.A., Lindrea K.C. (1989). A survey of filamentous bacterial populations from foaming activated sludge plants in eastern states of Australia. *Water Res*, **24**, 493-498.

Sezgin M., Lechevallier M.P., Karr P.R. (1988). Isolation and identification of actinomycetes present in activated sludge scum. *Water Sci & Technol*, **20**(11-12), 257-263.

Shao Y.J., Starr M., Kaporis K., Kim H.S., Jenkins D. (1997). Polymer addition as a solutions to *Nocardia* foaming Problems. *Water Environ Res*, **69**, 25-27.

Sladka A., Ottova V. (1973). Filamentous Organisms in Activated Sludge. *Hydrobiologia*, 43, 285–299.

Slijkhuis H. (1983). The physiology of the filamentous bacterium *Microthrix parvicella*. PhD thesis, Landbouwhogeschool, Wageningen, The Netherlands.

Slijkhuis H., Deinema M.H. (1988). Effect of environmental conditions on the occurrence of *Microthrix parvicella* in activated sludge. *Wat Res*, **22**, 825-828.

Snaidr J,, Beimfohr C., Levantesi C., Rossetti S., van der Waarde J., Geurkink B., Eikelboom D.H., Lemaitre M., Tandoi V, (2002). Phylogenetic analysis and in situ identification of *Nostocoida limicola*-like filamentous bacteria in activated sludge from industrial wastewater treatment plants. *Water Sci & Technol*, **46**(1-2), 99-104.

Soddell J.A., Seviour R.J. (1994). Incidence and morphological variability of *Nocardia pinensis* in Australian activated sludge plants. *Water Res*, **28**, 2343-2351.

Soddell J.A., Seviour R.J. (1998). Numerical Taxonomy of *Skermania piniformis* and related organisms from activated sludge. *J Appl Microbiol*, **84**, 272-284.

Soddell J.A. (1999). Foaming. In: "Microbiology of Activated Sludge", Seviour R.J., Blackall L.L. Eds., Kluwer Academic, London UK.

Soddell J.A., Seviour R.J., Blackall L.L., Hugenholtz P. (1998). New foam forming nocardioforms found in activated sludge. *Water Sci & Technol*, **37**(4-5), 495-502.

Starr M.A., Shao Y.J., Kaporis K., Kim H.S., Jenkins D. (1996). Polymer addition: a solution to N*ocardia* foaming problems at the Terminal Island treatment plant, Los Angles, California. *Proceedings of the Water Environment Federation, 69st Annual Conference & Exposition*, Wastewater Treatment Research and Municipal Wastewater Treatment, Volume 1, 183-188. Alexandria (VA) USA.

Still D.A.,Ekama G.A., Wentzel M.C., Casey T.G., Marais G.v.R. (1996). Filamentous organism bulking in nutrient removal activated sludge systems. Paper 2: Stimulation of the selector effect under aerobic conditions. *Water SA*, **22**, 97-118.

Stover & Associates Inc. *FIL-IDENTPro™ . The Computer-Based Rapid Filamentous Bacteria Identification Program*. Stillwater, OK, USA.

Stratton H.M., Seviour R.J., Soddell J.A., Blackall L.L., Muir D. (1996). The opportunistic pathogen *Nocardia farcinica* is a foam-producing bacterium in activated sludge plants. *Lett Appl Microbiol*, **22**, 341-346.

Strom P.F., Jenkins D. (1984). Identification and signification of filamentous microorganisms in activated sludge. *Wat Poll Cont*, **56**, 449-459.

Strom P.F., Jenkins D. (1984). Identification and significance of filamentous macroorganisms in activated sludge. *Journal of the Water Pollution Control Federation*, **56**, 449-59.

Tipping P. (1995). Foaming in activated-sludge processes: an operators overview. *J Instn Wat & Envir Mangt*, **9**, 281-289.

Tomlinson E., Chambers B. (1984). Control strategies for bulking sludge. *Water Sci & Technol*, **16**(10-11), 15-34.

van Leeuwen J. (1988). Bulking control with ozonation in a nutrient removal activated sludge plant. *Water SA*, **14**, 119-124.

van Leeuwen J. (1989). Ozonation for non-filamentous bulking control in an activated sludge plant treating fuel synthesis wastewater. *Water SA*, **15**, 127-132.

van Leeuwen J, (1990). Review of oxidative control of bulking in activated sludge waste treatment to improve sedimentation. *SA J of Chem Eng*, **2**, 27-40.

van Leeuwen J., Pretorius W.A. (1988). Sludge bulking control with ozone. *J IWEM*, **2**, 223-227 van Loosdrecht M.C.M., Pot M., Heijnen J.J. (1997). Importance of bacterial storage polymers in bioprocesses. *Water Sci & Technol*, **35**(1), 41-48 .

Vismara R. (1998). Problemi di sedimentabilità dei fanghi attivi. *Ingegneria Ambientale*, **11**, 414-427.

Wagner F. (1982). Study of the causes and prevention of sludge bulking in Germany. In: *Bulking of Activated Sludge: Preventive and Remedial Methods*. Chambers B., Tomlinson E.J. Eds., 29-41, Horwood Chichester.

Wagner M., Amann R., Kaemper P., Assmus B., Hartmann A., Hutzler P., Springer N., Schleifer K., (1994) Identification and in situ detection of Gram-negative filamentous bacteria in activated sludge. *System Appl Microbiol*, **17**, 405-417.

Wakefield R.W. (1998). Personal communication, *Port Elizabeth Municipality*.

Wanner J. (1994a). The Implementation of Bulking Control in the Design of Activated Sludge Plants *Water Sci & Tech*, **29**(7), 193–202.

Wanner J. (1994b). *Activated Sludge Bulking and Foaming Control*, Technomic Publishing, Co Lancaster, (PA) USA.

Wanner J., Grau P. (1989). Identification of filamentous microorganisms from activated sludge: a compromise between wishes needs and possibilities. *Wat Res*, **23**, 883–891.

Wanner J., Kucman K., Ottova V., Grau P. (1987). Effect of anaerobic conditions on activated sludge filamentous bulking in laboratory systems. *Wat Res*, **21**, 1541–1546.

Wanner J., Ruzickova I., Jetmarova P., Krhutkova O., Paraniakova J. (1998). A national survey of activated sludge separation problems in the Czech Republic: filaments floc characteristics and activated sludge metabolic properties. *Water Sci & Technol*, **37**(4-5), 271–279.

Wanner J., Ruzickova I., Krhutkova O., Pribyl M. (2000). Activated sludge population dynamics and wastewater treatment plant design and operation. *Water Sci & Technol*, **41**(9), 217–225.

Wanner J. (1994). The implementation of bulking control in designing activated sludge systems. *Water Sci & Technol*, **29**(7), 221-228.

Wanner J. (1994a). Activated sludge bulking and foaming, Technomic Publishing, Co Pen.

Wanner J. (1994b). Activated sludge population dynamics. *Water Sci & Technol*, **30**(11), 159-169.

Ward D.B., Oldham K., Abraham (1999). Resolution of capacity constraints and performance variations in a biological nutrient removal process. *Proceedings of the Water Environment Federation, 72nd Annual Conference & Exposition*, CD-Rom, Alexandria (VA), USA.

Yumikura J., Ishi M. (1998). Survey on effective separation of activated sludge. Annual Report of Japan Sewage Works Agency (in Japanese).

7

Modelling of bulking phenomena

M.C. Tomei

7.1 INTRODUCTION

The modelling of activated sludge bulking helps to better understanding this complex phenomenon by rapid and simple simulation of various operational situations. The calibrated models can be also used for the selection of proper control measures for a particular activated sludge system. Many studies of microbial selection mechanisms in biological reactors exist in the specialised literature: e.g. theoretical papers on population dynamics (i.e. Fredrickson and Stephanopoulos, 1981; Pavlou et al., 1990; Baltzis and Wu, 1994) and filamentous micro-organisms proliferation in activated sludge plants (Chudoba, et al., 1973; Chiesa and Irvine, 1985; Kappeler and Gujer, 1994a; Sheintuch and Tartakovsky, 1997). Identified causes of bulking are:

- influent characteristics (concentration and biodegradability of carbonaceous substrates, septicity, temperature, biomass inoculation, presence of toxic compounds),
- plant configuration (sequence of operations, reactor mixing conditions and compartmentalization)
- operating variables (hydraulic and sludge retention time, organic load, pH, COD, and oxygen and nutrient concentrations).

Some of the most important filamentous bacteria dominant in bulking have been isolated and physiologically characterized (Jenkins et al., 1993). Secondary settler dynamic modelling and interactions between the reactor and the settler have not been adequately considered in the past, mainly for the complexity of the system and the inadequacy of the available investigation techniques.

Significant improvements in bulking modelling are expected from techniques for *in situ* bacterial characterization (Amann et al., 1995), that allow a direct investigation into micro-organism behaviour in the reaction environment.

7.2 CHUDOBA'S THEORY

The first model for aerobic bulking was proposed by Chudoba et al. (1973) and is still a fundamental reference in this field. The model is based on kinetic competition between filamentous and floc-forming bacteria for a single soluble carbonaceous substrate over a range of substrate concentrations.

A Monod kinetic relationship between growth rate and substrate concentrations is used:

$$\frac{dX}{Xdt}=\mu=\mu_{max}\frac{S}{Ks+S} \tag{7.1}$$

where:
X=biomass concentration (ML^{-3})
t =time (T)
S= substrate concentration (ML^{-3})
Ks = saturation constant ($M_{substrate}L^{-3}$)
μ_{max} = maximum specific growth rate (T^{-1})
μ = specific growth rate (T^{-1})

The ratio μ_{max}/Ks is the key factor in determining the dominant population. Floc-forming bacteria have the higher values of μ_{max}, and Ks. At low substrate concentrations the low Ks value of the filaments (corresponding to a high substrate affinity) makes the μ_{max}/Ks ratio (and consequently the growth rate) higher for the filaments (Ks strategists) than for the floc-formers. The opposite situation occurs at high substrate concentrations where the Ks effect is negligible and the higher μ_{max} values favour the floc-formers (μ_{max} strategists).

Chudoba's model predicts that filamentous micro-organisms are favoured in completely mixed reactors (CSTR), operating at high substrate removal efficiency (low substrate concentration). On the contrary, floc formers will be favoured in plug flow reactors or, more generally, in plant configurations operating at high substrate concentrations in the inlet zone. These predictions help to explain the effect of bulking control with plug flow and compartmentalized basins reactors compared to CSTR and the beneficial effect of selectors (see chapter 5).

The differences between the kinetic parameters (Table 7.1) of the two groups of micro-organisms are not large, (Wanner, 1993). The μ_{max}/Ks ratio is only significantly different if extreme values are compared. Because municipal wastewater is a complex mixture of different compounds varying in proportion and composition (see chap. 1), it is difficult to reliably determine μ_{max} and Ks values. A further uncertainty occurs: in flocculent cultures these parameters accounts for both the intrinsic kinetics and diffusional resistance. The latter is negligible for dispersed pure cultures but can be significant for the floc sizes usually found in activated sludge plants (Beccari et al., 1992). The "apparent" Ks value is not specific for the micro-organisms substrate concentration but is strongly dependent on operating plant characteristics.

In conclusion, the simple competitive approach proposed can be successfully applied to describe bulking where the filament responsible has kinetic parameters that significantly differ from floc-former values.

Table 7.1 Summary of kinetic parameter literature data for the characteristic representatives of floc-forming and filamentous groups (from Wanner, 1993). Reference substrates: acetate and glucose.

Micro-organism group	μ_{max} (d^{-1})	Ks (mg COD l^{-1})
Floc-formers (*Citrobacter spp., Zooglea ramigera*)	1 - 9.2	0.3 - 5
Filaments (*Sphaerotilus natans, Type 1701, Type 021N, Haliscomenobacter hydrossis, Nocardia sp.*)	1.2 - 6.5	1 - 10

7.3 CHIESA AND IRVINE MODEL

Chiesa and Irvine (1985) proposed a more complex description of bacterial competition in activated sludge, based on a literature review of the physiology of micro-organisms most frequently detected in activated sludge. The micro-organisms were characterized by substrate affinity, resistance to starvation and response to DO concentration (Table 7.2).

Table 7.2 Physiological characteristics of the main bacterial groups found in activated sludge (from Chiesa and Irvine, 1985)

Traits	Common non filamentous bacteria	Filaments associated with high organic loading rates	Filaments associated with low organic loading rates
Substrate affinity	Low	High	High
Maximum growth rate	High	High	Low
Endogenous metabolic rate	Low	High	Low
Resistance to starvation	High	Low	Very high
Potential to accumulate storage compounds	High	High	Varies
Maximum rate of substrate uptake/storage product formation	Very high	High	Low
Decrease in metabolic activity at low dissolved oxygen	High	Slight	-

The idealized activated sludge biomass consisted of three model organisms competing for one soluble substrate:

- fast growing zoogleal (floc forming) bacteria with high starvation resistance and whose metabolic activity decreases at a low DO concentration (I)
- slow growing starvation resistant filaments with a high substrate affinity (low Ks value) (II)
- fast growing starvation susceptible filaments with a high affinity for DO (resistant to low DO levels) (III).

There is a substrate concentration value, S*, at which the three groups coexist, at high DO concentrations. For $S < S^*$ the slow growing starvation resistant filament will prevail, while for $S > S^*$ the floc forming zoogleal micro-organisms will prevail. At low DO levels three intervals of carbonaceous substrate concentration, corresponding to the prevalence of different groups, exist:

- $S < S^*$ where group II is dominant;
- $S^* < S < S^{**}$ where group I (zoogloeal organisms) prevails
- $S>S^{**}$ where group III is dominant.

The most important feature of the Chiesa and Irvine model is the detailed characterization of filamentous groups, that accounts their behaviour under different conditions of substrate concentration, DO concentration and the absence of exogenous substrate.

7.4 AEROFIL MODEL

7.4.1 Hypotheses and fundamentals

A more complete description of bulking is presented in the "aerobic bulking model", AEROFIL, proposed by Kappeler and Gujer (1994a, b). This simulates a mixed culture constituted of facultative aerobic floc-forming, obligate aerobic filamentous and nitrifying micro-organisms in plants that include anaerobic, anoxic and aerobic stages.

The acronym AEROFIL derives from "obligate AERObic FILamentous, the bacterial group that, the Authors propose is responsible for aerobic bulking in the majority of cases and includes *Sphaerotilus natans*, Type 021N and Type 0961.

The model is restricted to the following cases:

- completely aerobic systems operating at SRT <10 days;
- aerobic systems with small anoxic zones (selector) and SRT < 10 days;
- aerobic systems with a small anaerobic zone (selector) a SRT so that does not allow EBPR.
- low surfactant levels (favouring the proliferation of *Actinomycetes*) and low long chain fatty acids (favouring the proliferation of *Michrothrix parvicella*) whose metabolism is not predicted according the usual kinetic equations.

Distribution of floc-forming and filamentous micro-organisms in the reactor is determined by competition for the readily biodegradable COD. Prevalence also depends on physical factors such as the biofloc size of the and diffusional resistances in the solid and liquid phases.

According to the Activated Sludge Model (ASM) No. 1 (Henze et al., 1987), biodegradable COD influent is constituted by a soluble readily biodegradable fraction (S_S) and a particulate and colloidal slowly biodegradable fraction (X_S) (see Chap. 1). The X_S fraction has to be hydrolysed by floc-formers before it can be utilised by both species. Hydrolysis products (S_H) are similar to readily biodegradable COD, therefore two components of readily biodegradable COD are defined: S_S, present in the influent, and S_H), originating from hydrolysis. These two components are utilized by the two populations with different kinetics. S_S is immediately available for filaments which reach out of the flocs in the bulk solution but is not easily utilizable by the floc-formers because it must diffuse into the flocs prior to being metabolised. The opposite situation exists for S_H, where hydrolysis of particulate and colloidal organic material (X_S) takes place within the activated sludge flocs and hydrolysis products (S_H) are primarily available to the floc-formers. Diffusion from the inside the floc to the external bulk phase is necessary to make S_H available to filaments. Thus the relative abundance of S_S and X_S (hydrolysed to S_H) can give a selective advantages to filaments or floc-formers.

The simplest way to model the presence of a diffusional stage is to use the Monod kinetics for the entire process (diffusion plus biological reaction) and increase Ks saturation constant above its intrinsic value. The "apparent" Ks value simulates the effect of diffusional resistance.

The Authors use a low Ks value for hydrolysis product utilization by flocforming organisms and a high value for S_S. The opposite situation is assumed for filamentous organisms with a low Ks value for S_S and a high Ks value for S_H.

The COD fractions S_I and X_I do not have any effect on the microbial competition.

7.4.2 Processes in model

The structure of the AEROFIL model is similar to that of the ASM1 (Henze et al., 1987) and ASM2 (Henze et al., 1995): table 7.3 reports kinetic parameters, their definitions and the proposed default values for the three populations constituting the biocenosis, while tables 7.4 and 7.5 report the kinetic equations and the stoichiometry of the processes included in the model. Table 7.4 shows that the rather complex kinetic equations are first order with biomass concentration and shifting order with limiting substrates. In addition, inhibition effect is described with a specific "switching function".

The effect of limiting substrates (COD, ammonia-N, nitrate-N, oxygen) is taken into account by utilizing the usual variable order kinetics, in which the maximum reaction rate value is reduced by a factor given by:

$$\frac{S}{Ks+S} \tag{7.2}$$

where S is the substrate concentration and Ks the half saturation constant: each substrate has a reducing factor.

The inhibition effect of some components is considered in the kinetic equations by introducing "switching functions" defined by:

$$\frac{K_I}{K_I+S} \tag{7.3}$$

where S is the inhibitory compound concentration and K_I is the inhibition constant.

Differences between the AEROFIL model and with the ASM1 and ASM2 models are in their treatment of decay, hydrolysis and nitrification/denitrification.

Table 7.3. Kinetic parameters: definition and default values (20 °C) (from Kappeler and Gujer, 1994a)

Kinetic parameters		X_{Floc}	X_{Nitr}	X_{Fil}	Units
Maximum specific growth rate	μ_{max}	3.0	0.9	3.0	d^{-1}
Half-saturation coefficient for S_S	K_s	8	-	1	g COD m^{-3}
Half-saturation coefficient for S_H	K_H	1	-	4	g COD m^{-3}
Half-saturation coefficient for O_2	K_{O2}	0.25	0.4	0.1	$gO_2\ m^{-3}$
Half-saturation coefficient for NO_3	K_{NO3}	0.5	-	-	gNO_3-N m^{-3}
Coefficient for reduced anoxic growth	η_g	0.8	-	-	-
Half-saturation coefficient for NH_4	K_{NH4}	0.01	0.4	0.01	gNH_4-N m^{-3}
Yield coefficient of X_{Floc} and X_{Fil}	Y_H	0.63	-	0.63	gCOD $gCOD^{-1}$
Yield coefficient of X_{Nitr}	Y_A	-	0.24		gCOD gN^{-1}
Coefficient for lysis	b'	0.5	0.14	0.5	d^{-1}
Fractional coefficient for the release of inert particulate COD by lysis	f_I	0.08	0.08	0.08	gCOD $gCOD^{-1}$
Correction factor for endogenous respiration	η_{endog}	0.4	0.4	0.4	-
Fractional coefficient for the release of inert particulate COD by endogenous respiration	f_p	0.2	0.2	0.2	gCOD $gCOD^{-1}$
Ratio of nitrogen to COD in biomass and inert particulate COD	i_{XB}	0.06	0.06	0.06	gN $gCOD^{-1}$
Hydrolysis constant under aerobic conditions	k_h		5		d^{-1}
Fractional coefficient for the release of soluble inert COD by hydrolysis	f_H		0.07		gCOD $gCOD^{-1}$
Coefficient for reduced anoxic hydrolysis	η_h		0.6		-
Fraction of released ammonium by hydrolysis	i_{XS}		0.04		gN $gCOD^{-1}$

Table 7.4 Process kinetics of the AEROFIL model (units g m^{-3}d^{-1}) (from Kappeler and Gujer, 1994a)

Process	Process rate
Facultative aerobic floc-formers X_{Floc}	
Aerobic growth S_S	$\mu_{max,Floc}\cdot\left(\frac{S_S}{K_s+S_S}\cdot\frac{S_S}{S_S+S_H}\right)\cdot\left(\frac{S_{NH_4}}{K_{NH_4}+S_{NH_4}}\right)\cdot\left(\frac{S_{O_2}}{K_{O_2}+S_{O_2}}\right)\cdot X_{Floc}$
Aerobic growth S_H	$\mu_{max,Floc}\cdot\left(\frac{S_H}{K_H+S_H}\cdot\frac{S_H}{S_S+S_H}\right)\cdot\left(\frac{S_{NH_4}}{K_{NH_4}+S_{NH_4}}\right)\cdot\left(\frac{S_{O_2}}{K_{O_2}+S_{O_2}}\right)\cdot X_{Floc}$
Anoxic growth S_S	$\eta_g\cdot\mu_{max,Floc}\cdot\left(\frac{S_S}{K_S+S_S}\cdot\frac{S_S}{S_S+S_H}\right)\cdot\left(\frac{S_{NO_3}}{K_{NO_3}+S_{NO_3}}\right)\cdot\left(\frac{S_{NH_4}}{K_{NH_4}+S_{NH_4}}\right)\cdot\left(\frac{K_{O_2}}{K_{O_2}+S_{O_2}}\right)\cdot X_{Floc}$
Anoxic growth S_H	$\eta_g\cdot\mu_{max,Floc}\cdot\left(\frac{S_H}{K_H+S_H}\cdot\frac{S_H}{S_s+S_H}\right)\cdot\left(\frac{S_{NO_3}}{K_{NO_3}+S_{NO_3}}\right)\cdot\left(\frac{S_{NH_4}}{K_{NH_4}+S_{NH_4}}\right)\cdot\left(\frac{K_{O_2}}{K_{O_2}+S_{O_2}}\right)\cdot X_{Floc}$
Endogenous respiration	$\eta_{endog}\cdot b'_{Floc}X_{Floc}\cdot\left(\frac{S_{O_2}}{K_{O_2}+S_{O_2}}\right)$
Lysis	$b'_{Floc}\cdot X_{Floc}\cdot\left(\frac{K_{O_2}}{K_{O_2}+S_{O_2}}\right)$

Table 7.4 (cont'd)

Nitrifiers X_{Nit}	
Growth	$\mu_{max,Nitr} \cdot \left(\frac{S_{O_2}}{K_{O_2}+S_{O_2}} \frac{S_{NH_4}}{K_{NH_4}+S_{NH_4}} \right) \cdot X_{Nitr}$
Endogenous respiration	$\eta_{endog} \cdot b'_{Nitr} X_{Nitr} \cdot \left(\frac{S_{O_2}}{K_{O_2}+S_{O_2}} \right)$
Lysis	$b'_{Nitr} X_{Nitr} \cdot \left(\frac{K_{O_2}}{K_{O_2}+K_{O_2}} \right)$
Obligate aerobic filaments X_{Fil}	
Aerobic growth S_S	$\mu_{max,Fil} \cdot \left(\frac{S_S}{K_s+S_S} \cdot \frac{S_S}{S_S+S_H} \right) \cdot \left(\frac{S_{NH_4}}{K_{NH_4}+S_{NH_4}} \right) \cdot \left(\frac{S_{O_2}}{K_{O_2}+S_{O_2}} \right) \cdot X_{Fil}$
Aerobic growth S_H	$\mu_{max,Fil} \cdot \left(\frac{S_H}{K_H+S_H} \cdot \frac{S_H}{S_S+S_H} \right) \cdot \left(\frac{S_{NH_4}}{K_{NH_4}+S_{NH_4}} \right) \cdot \left(\frac{S_{O_2}}{K_{O_2}+S_{O_2}} \right) \cdot X_{Fil}$
Endogenous respiration	$\eta_{endog}\ b'_{Fil} \cdot X_{Fil} \cdot \left(\frac{S_{O_2}}{K_{O_2}+S_{O_2}} \right)$
Lysis	$b'_{Fil} \cdot X_{Fil} \cdot \left(\frac{K_{O_2}}{K_{O_2}+S_{O_2}} \right)$
Hydrolysis	
Aerobic	$k_h \cdot X_S \cdot \left(\frac{S_{O_2}}{K_{O_2}+S_{O_2}} \right)$
Anoxic	$\eta_h \cdot k_h \cdot X_S \cdot \left(\frac{K_{O_2}}{K_{O_2}+S_{O_2}} \right) \cdot \left(\frac{S_{NO_3}}{K_{NO_3}+S_{NO_3}} \right)$

Table 7.5. Stoichiometry of the AEROFIL model (from Kappeler and Gujer,1994a)

Process	S_{O2}	S_S	S_H	S_I	X_S	X_I	X_{Floc}	X_{Nitr}	X_{Fil}	S_{NH4}	S_{NO3}
Facultative aerobic floc-formers X_{Floc}											
Aerobic growth on S_S	$-\frac{1-Y_H}{Y_H}$	$-\frac{1}{Y_H}$					+1			$-i_{XB}$	
Aerobic growth on S_H	$-\frac{1-Y_H}{Y_H}$		$-\frac{1}{Y_H}$				+1			$-i_{XB}$	
Anoxic growth on S_S		$-\frac{1}{Y_H}$					+1			$-i_{XB}$	$-\frac{1-Y_H}{2.86Y_H}$
Anoxic growth on S_H			$-\frac{1}{Y_H}$				+1			$-i_{XB}$	$-\frac{1-Y_H}{2.86Y_H}$
Endogenous respiration	$-(1-f_p)$					$+f_p$	-1			$i_{XB}\cdot(1-f_p)$	
Lysis			$1-f_I$			$+f_I$	-1			$i_{XB}\cdot(1-f_I)$	
Nitrifies X_{Nitr}											
Growth	$-\frac{4.57-Y_A}{Y_A}$							+1		$-i_{XB}-\frac{1}{Y_A}$	$-\frac{1}{Y_A}$
Endogenous respiration	$-(1-f_p)$					$+f_p$		-1		$i_{XB}\cdot(1-f_p)$	
Lysis			$1-f_I$			$+f_I$		-1		$i_{XB}\cdot(1-f_I)$	

Table 7.5 (cont'd).

Process	S_{O2}	S_S	S_H	S_I	X_S	X_I	X_{Floc}	X_{Nitr}	X_{Fil}	S_{NH4}	S_{NO3}
Obligate aerobic filaments X_{Fil}											
Aerobic growth on S_S	$-\frac{1-Y_H}{Y_H}$	$-\frac{1}{Y_H}$							+1	$-i_{XB}$	
Aerobic growth on S_H	$-\frac{1-Y_H}{Y_H}$		$-\frac{1}{Y_H}$						+1	$-i_{XB}$	
Endogenous respiration	$-(1-f_p)$					$+f_p$			-1	$i_{XB} \cdot (1-f_p)$	
Lysis		$1-f_I$				$+f_I$			-1	$i_{XB} \cdot (1-f_I)$	
Hydrolysis											
Aerobic			$1-f_H$	f_H	-1					i_{XS}	
Anoxic			$1-f_H$	f_H	-1					i_{XS}	
Units	COD	COD	COD	COD	COD	COD	COD	COD	COD	N	N

Biomass decay rate depends on DO; decay processes for bacteria groups of are described as endogenous respiration under aerobic conditions (with a limiting term depending on the DO concentration) and as lysis when oxygen is absent (with a switching function depends on the DO concentration).

Hydrolysis process kinetics are simplified by using the conclusion of Wanner et al. (1992), that complex mathematical descriptions do not improve interpretation of the experimental data. Simple first order kinetics relative to X_S are utilized and it is assumed that the process takes place only under aerobic and anoxic conditions but not under anaerobic zones; the kinetics under aerobic conditions has a limiting factor depending on the DO concentration while in anoxic conditions a limiting factor depending on N- NO_3 concentration and a switching function depending on DO concentration are used.

Nitrification and denitrification process kinetics are described with simple equations and only N- NH_3 and N- NO_3 are considered as nitrogen compounds.

Facultative aerobic floc-formers (X_{Floc}) growth is considered under both aerobic and anoxic conditions for the two readily biodegradable COD fractions S_S and S_H. Aerobic growth kinetics include limiting terms for the carbonaceous substrates, the N- NH_3 and the DO concentration. In anoxic conditions the same approach is followed with a limiting term for N- NO_3 and a switching function depending on the DO concentration.

Obligate aerobic filaments X_{Fil} can grow only under aerobic conditions and the kinetic equations are the same as for floc- formers.

Nitrifiers (X_{Nitr}) have a double substrate growth equation relative to N- NH_3 and DO concentrations.

The default parameters, (Table 7.3) are based on the Author's experience. Particular attention has to be given to the "apparent" Ks value, the key parameter in determining the floc composition. It is obtained from (Charaklis, 1978):

$$Ks = Ks_0 + \left(\frac{\mu_{max} \gamma \delta}{Y_H a D_S} \right) \tag{7.4}$$

where:
Ks_0 = intrinsic substrate saturation coefficient (ML^{-3})
μ_{max} = maximum specific growth rate (T^{-1})
γ = biomass density (ML^{-3})
δ = diffusional length (L)
Y_H = growth yield coefficient (M biomass M substrate^{-1})
a = specific surface of biological solids (L^{-1})
D_S = diffusion coefficient of substrate (L^2T^{-1})

The defined parameters can be determined by model calibration thereby accounting for the specific characteristics of the system under consideration.

The AEROFIL model has been successfully tested on full-scale and pilot scale plants. It has allowed verification of the effects on the sludge composition of important parameters such as influent inoculum, plant configuration, organic load and readily biodegradable COD influent fractions. It has been used to evaluate the optimum selector (both aerobic and anoxic) volumes. Details of model validation are reported in Kappeler and Gujer, 1994b.

7.5 CONCLUSIONS

The complexity of bulking derives both from the variety of causes and the lack of information on the filamentous bacteria involved. This currently makes formulation of a deterministic model of general validity difficult. It is expected that "in the near future" recent improvements in microbiological techniques for *in situ* bacterial characterization will provide more reliable information on the population dynamics under operating conditions. Reaching this objective is strongly dependent on the integration of engineering and microbial investigations.

An interesting alternative to deterministic models, particularly useful in control applications, are expert systems that utilize empirical information rather than mathematical equations as database.

7.6 REFERENCES

Amann R.I., Ludwig W., Schleifer K. (1995). Phylogenetic identification and in situ detection of individual microbial cells without cultivation. *Microbiol. Rev.*, **59**, 143-169.

Baltzis B.C., Wu Ming (1994). Steady state coexistence of three pure and simple competitors in a four membered reactor network. *Mathematical Biosciences*, **123**, 147-165.

Beccari M., Di Pinto A.C., Ramadori R., Tomei M.C. (1992). Effect of DO and diffusional resistances on nitrification kinetics. *Water Res*, **26**, 1099-1104.

Charaklis W.G. (1978). Microbial reaction rate expressions. *J. Envir. Engng. Div. Am. Soc. Civ. Engrs* **EE3** 531-534.

Chiesa S.C., Irvine R.L. (1985). Growth and Control of filamentous microbes in activated sludge: an integrated hypothesis. *Water Res*, **19**, 471-479.

Chudoba J., Grau P., Ottova V. (1973). Control of activated sludge filamentous bulking II. Selection of micro-organisms by means of a selector. *Water Res*, **7**, 1389-1406.

Friedrickson A.G., Stephanopoulos G. (1981). Microbial Competition. *Science*, **213**, 972-979.

Henze M., Grady Jr. C.P.L., Gujer W., Marais G.v.R., Matsuo T. (1987). Activated Sludge Model No 1. *IAWPRC Scientific and Technical Reports No 1*, IAWPRC London, UK.

Henze M., Gujer W., Mino T., Matsuo T., Wentzel M.C., Marais G.v.R. (1995). Activated Sludge Model No 2. *IAWQ Scientific and Technical Reports No 2*, IAWQ London, UK.

Jenkins D., Richard M.G., Daigger M.R. (1993). Manual on the causes and control of activated sludge bulking and foaming. 2nd Edition, Lewis Publisher, (MI) USA.

Kappeler J., Gujer W. (1994a). Development of a mathematical model for "aerobic bulking". *Water Res*, **28**, 303-310.

Kappeler J., Gujer W.(1994b). Verification and applications of a mathematical model for "aerobic bulking". *Water Res*. **28**, 311-322.

Pavlou S., Kevrekidis I.G., Lyberatos G. (1990). On the coexistence of competing microbial species in a chemostat under cycling. *Biotechnol Bioeng*, **35**, 224-232.

Sheintuch M., Tartakovsky B. (1997). Activated-sludge system design for species selection: analysis of a detailed multispecies model. *Chem Eng Sci*, **52**, 3033-3046.

Wanner J.(1993). Kinetic and metabolic selection in controlling the filamentous organisms in activated sludge systems. In *Proceedings of the International Workshop on Prevention and Control of Bulking Activated Sludge*. Jenkins D., Ramadori R., Cingolani L. Eds. Perugia, 22-23 June 1992.

Wanner O., Kappeler J., Gujer W. (1992). Calibration of an activated sludge model based on human expertise and on mathematical optimization technique- a comparison. *Wat Sci & Techol,* **25** (6) 141-148.

Index

Note: page numbers in *italics* refer to Figures while those in **bold** denote Tables.

IWA Publishing's authorised EU representative for General Product Safety Regulations is Diane D'Arras, 15 rue Duret, 75116 Paris, France, e-mail: safety@iwap.co.uk.

Printed and bound by CPI Group (UK) Ltd, Croydon, CR0 4YY

12/05/2026

02108909-0003